Chromatographic Separations

Analytical Chemistry by Open Learning

Titles in Series:

Chromatographic Separations

Analytical Chemistry by Open Learning

Authors:
PETER A SEWELL
Liverpool Polytechnic

BRIAN CLARKE
Neath College

Editor:
DAVID KEALEY

on behalf of ACOL

Published on behalf of ACOL, Thames Polytechnic, London
by
JOHN WILEY & SONS
Chichester · New York · Brisbane · Toronto · Singapore

Published by permission of the Controller of
Her Majesty's Stationery Office

Library of Congress Cataloging in Publication Data:

Sewell, P. A. (Peter Alexis)
  Chromatographic separations.
  (Analytical Chemistry by Open Learning)
  Bibliography: p.
  1. Chromatographic analysis—Programmed instruction. 2. Chemistry,
Analytic—Programmed instruction.   I. Clarke, Brian P. II. Kealey, D. (David)
III. Title. IV. Series: Analytical Chemistry by Open Learning (Series).
QD79.C4S49   1987   543'.089'077   87-13806

ISBN 0 471 91370 7
ISBN 0 471 91371 5 (Pbk.)

British Library Cataloguing in Publication Data:

Sewell, P. A.
  Chromatographic separations.—(Analytical chemistry).
  1. Chromatographic analysis I. Title   II. Clarke, Brian   III. Kealey, D

  IV. ACOL V. Series
  543'.089 QD79.C4

ISBN 0 471 91370 7
ISBN 0 471 91371 5 (Pbk)

Printed and bound in Great Britain

Analytical Chemistry

This series of texts is a result of an initiative by the Committee of Heads of Polytechnic Chemistry Departments in the United Kingdom. A project team based at Thames Polytechnic using funds available from the Manpower Services Commission 'Open Tech' Project has organised and managed the development of the material suitable for use by 'Distance Learners'. The contents of the various units have been identified, planned and written almost exclusively by groups of polytechnic staff, who are both expert in the subject area and are currently teaching in analytical chemistry.

The texts are for those interested in the basics of analytical chemistry and instrumental techniques who wish to study in a more flexible way than traditional institute attendance or to augment such attendance. A series of these units may be used by those undertaking courses leading to BTEC (levels IV and V), Royal Society of Chemistry (Certificates of Applied Chemistry) or other qualifications. The level is thus that of Senior Technician.

It is emphasised however that whilst the theoretical aspects of analytical chemistry can be studied in this way there is no substitute for the laboratory to learn the associated practical skills. In the U.K. there are nominated Polytechnics, Colleges and other Institutions who offer tutorial and practical support to achieve the practical objectives identified within each text. It is expected that many institutions worldwide will also provide such support.

The project will continue at Thames Polytechnic to support these 'Open Learning Texts', to continually refresh and update the material and to extend its coverage.

Further information about nominated support centres, the material or open learning techniques may be obtained from the project office at Thames Polytechnic, ACOL, Wellington St., Woolwich, London, SE18 6PF.

How to Use an Open Learning Text

Open learning texts are designed as a convenient and flexible way of studying for people who, for a variety of reasons cannot use conventional education courses. You will learn from this text the principles of one subject in Analytical Chemistry, but only by putting this knowledge into practice, under professional supervision, will you gain a full understanding of the analytical techniques described.

To achieve the full benefit from an open learning text you need to plan your place and time of study.

- Find the most suitable place to study where you can work without disturbance.

- If you have a tutor supervising your study discuss with him, or her, the date by which you should have completed this text.

- Some people study perfectly well in irregular bursts, however most students find that setting aside a certain number of hours each day is the most satisfactory method. It is for you to decide which pattern of study suits you best.

- If you decide to study for several hours at once, take short breaks of five or ten minutes every half hour or so. You will find that this method maintains a higher overall level of concentration.

Before you begin a detailed reading of the text, familiarise yourself with the general layout of the material. Have a look at the course contents list at the front of the book and flip through the pages to get a general impression of the way the subject is dealt with. You will find that there is space on the pages to make comments alongside the text as you study—your own notes for highlighting points that you feel are particularly important. Indicate in the margin the points you would like to discuss further with a tutor or fellow student. When you come to revise, these personal study notes will be very useful.

∏   When you find a paragraph in the text marked with a symbol such as is shown here, this is where you get involved. At this point you are directed to do things: draw graphs, answer questions, perform calculations, etc. Do make an attempt at these activities. If necessary cover the succeeding response with a piece of paper until you are ready to read on. This is an opportunity for you to learn by participating in the subject and although the text continues by discussing your response, there is no better way to learn than by working things out for yourself.

We have introduced self assessment questions (SAQ) at appropriate places in the text. These SAQs provide for you a way of finding out if you understand what you have just been studying. There is space on the page for your answer and for any comments you want to add after reading the author's response. You will find the author's response to each SAQ at the end of the text. Compare what you have written with the response provided and read the discussion and advice.

At intervals in the text you will find a Summary and List of Objectives. The Summary will emphasise the important points covered by the material you have just read and the Objectives will give you a checklist of tasks you should then be able to achieve.

You can revise the Unit, perhaps for a formal examination, by re-reading the Summary and the Objectives, and by working through some of the SAQs. This should quickly alert you to areas of the text that need further study.

At the end of the book you will find for reference lists of commonly used scientific symbols and values, units of measurement and also a periodic table.

Contents

Study Guide

This Unit is intended to introduce you to the technique of chromatography. The Unit contains information on the theory and the practice that is the common basis for the majority of chromatographic techniques. The Unit is intended to provide you with the vocabulary that will help you in your subsequent study of techniques such as thin-layer, gas and high peformance liquid chromatography.

We will assume that you have an understanding of chemistry, equivalent to that of a student who has gained a Higher National Certificate in Chemistry. There are five Parts in this Unit, each dealing with a specific topic of chromatography.

Parts 1–3 introduce you to the principles of chromatographic separations, the underlying theory and how it relates to practical chromatography.

Part 4 covers the use of chromatography in qualitative and quantitative analysis.

The final Part deals with practical aspects associated with classical column chromatography and acts as an introduction to modern liquid chromatography.

Study of the detailed mathematical treatments of retention equations (Section 2.1.1) and band broadening processes (Section 3.5) can be regarded as optional. However, it is important to understand the practical significance and uses of the derived equations in the various forms of chromatography.

While we have tried to make the contents as clear as possible, there may be times when you wish to refer to additional sources of information. Suggested texts are listed in the following Bibliography.

Supporting Practical Work

1. GENERAL CONSIDERATIONS

Most laboratories will have some equipment for chromatography though in some cases this may be only the most basic, for thin layer and column chromatography. However, for an understanding of the principles of separation by chromatography this simple equipment may be adequate. For laboratories with more than the basic equipment some more advanced exercises are suggested.

2. AIMS

The principal aims are:

(*i*) To provide basic experience in handling the equipment for chromatography.

(*ii*) To illustrate important principles from the theory.

(*iii*) To enable the student to relate theory to practice.

3. EXPERIMENTS

(1) Separation of dyestuffs using normal phase tlc.

(2) Optimisation of a separation using both normal phase and reverse phase tlc.

(3) Ion-exchange column chromatography of a mixture of metal cations or biological materials.

(4) Separation of a mixture of three substances of different polarity

by column chromatography (*J Chem Ed 61*, 1019, 1984).

(5) Separation of proteins by size exclusion chromatography (*J Chem Ed 61*, 1021, 1984).

The following experiments could be done if a simple gas chromatograph is available.

(6) Plotting a van Deemter curve and determination of the constant terms A, B and C.

(7) Separation of substances of different polarity (eg cyclohexane, benzene, propanone, ethylethanoate and ethanol) on a non-polar phase (Apiezon L) and on a polar phase (Carbowax 20M). Measurement of the resolution of cyclohexane and benzene on the two columns.

Further experiments are contained in the following textbooks:

(*i*) D T Sawyer, W R Heinemann and J M Biebe, *Chemistry Experiments for Instrumental Analysis*, John Wiley and Sons, 1984.

(*ii*) J H Knox (Editor), *High Performance Liquid Chromatography*, Edinburgh University Press.

(*iii*) A Braithwaite and F J Smith, *Chromatographic Methods* (4th Edition), Chapman and Hall, 1985.

Bibliography

1. 'STANDARD' ANALYTICAL CHEMISTRY TEXTBOOKS

All textbooks on instrumental methods of analysis contain chapters on chromatography, and these make a useful introduction to the subject. Examples are:

(a) F W Fifield and D Kealey, *Principles and Practice of Analytical Chemistry*, International Textbook Co Ltd, 2nd Edn, 1983.

(b) D A Skoog and D M West, *Fundamentals of Analytical Chemistry*, 4th Edition, Holt-Saunders International Editions, 1976.

(c) H H Willard, L L Merritt, J A Dean and F A Settle, *Instrumental Methods of Analysis*, Van Nostrand, 1981.

2. CHROMATOGRAPHY TEXTBOOKS

There are specialist texts on almost all aspects of chromatography, but of a more general nature the following are recommended:

(a) A Braithwaite and F J Smith, *Chromatographic Methods*, 4th Edition, Chapman and Hall, 1985.

(b) R J Hamilton and P A Sewell, *Introduction to High Performance Liquid Chromatography*, 2nd Edition, Chapman and Hall, 1982.

(c) C F Poole and S A Schuette, *Contemporary Practice of Chromatography*, Elsevier Science Publishers, 1981.

Acknowledgements

Figures 3.3i and 3.3j are redrawn from *Journal of Physical Chemistry*, **75** (25) 3870, 1971 with the permission of the American Chemical Society.

Figures 3.6d, 3.6g and 3.6h are redrawn from L. R. Snyder, *Journal of Chromatographic Science*, **10**, 200–212, 1972, by permission of Preston Publications, a division of Preston Industries Inc.

Figure 4.2c is redrawn from R. E. Majors, *Journal of Chromatographic Science*, **8**, 338, 1970 by permission of the author and Preston Publications, a division of Preston Industries Inc.

Figure 4.3c is redrawn from advertising material for Magnus Data, Aylesbury with permission by Nelson Analytical, Wirral, Merseyside.

Figure 5.1a is redrawn from V. Heines, *Chem Tech*, 280, 1971. Permission has been requested.

Figure 5.2b is redrawn from Snyder and Kirkland, *Introduction to Modern Liquid Chromatography*, 2nd ed, John Wiley, 1979 with permission by John Wiley & Sons Inc.

Figure 5.2o is redrawn from the Chrompack General Catalogue, 1986. Permission has been requested.

Figure 5.3b is redrawn from K. A. Kraus and F. Nelson, *Proceedings of the First United Nations Conference on the Peaceful Uses of Atomic Energy*, 7, 113, 1956 by permission from the United Nations Publications Board.

Figure 5.3c is taken from Dionex advertising material, with permission of the Dionex Corporation.

Figure 5.3d is redrawn from Waters advertising material, with permission of Millipore–Waters Chromatography Division.

Figure 5.3e is redrawn from P. J. Twitchett, A. E. P. Gorvin and A. C. Moffat, *Journal of Chromatography*, **120**, 359, 1976 by permission of Elsevier Scientific Publishing Company, Amsterdam.

1. Introduction

Overview

This section introduces some of the language of chromatography, classifies chromatographic methods according to technique, methods of development and mode of separation and introduces the underlying physical chemical principles which account for the retention of sample molecules in a chromatographic system.

1.1. CHROMATOGRAPHY

Chromatography [literally 'colour-writing' from the Greek] is a separative technique whereby the components of a mixture may be separated by allowing the sample (*analyte*) to be transported through a packed bed of material (the *stationary phase*) by a fluid *mobile phase*. If the individual components in the mixture move through the packed bed at different rates then separation will occur, the degree of separation depending on the difference in the rates of migration.

The International Union of Pure and Applied Chemistry (IUPAC) has defined chromatography as:

'A method used primarily for the separation of components of a sample, in which the components are distributed between two phases, one of which is stationary while the other moves. The stationary phase may be a solid, or a liquid supported on a solid, or a gel. The stationary phase may be packed in a column, spread as a layer, or distributed as a film, etc. In these definitions, "chromatographic bed" is used as a general term to denote any of the different forms in which the stationary phase may be used. The mobile phase may be gaseous or liquid'

[*Recommendations on Nomenclature for Chromatography Rules Approved 1973*, IUPAC Analytical Chemistry Division Commission on Analytical Nomenclature, *Pure and Applied Chemistry 37*, 447, 1974.]

∏   What would you consider to be the essential features of a chromatographic system?

       From the IUPAC definition we would conclude that the essential features are two phases, one of which is stationary and the other is moving or mobile.

If the fluid mobile phase is a gas (called the *carrier* gas) the technique is known as *Gas Chromatography*; if it is a liquid then the technique is called *Liquid Chromatography*. Each technique may be further sub-divided according to the nature of the stationary phase, thus:

Nature of Mobile Phase	Nature of Stationary Phase	Name
Gas	Solid	Gas-solid chromatograpy
	Liquid	Gas-liquid chromatography
Liquid	Solid	Liquid-solid chromatography
	Liquid	Liquid-liquid chromatography

The stationary phase may be in the form of a flat bed of material consisting of an adsorbed layer spread evenly over a sheet of glass, plastic or aluminium (called *thin-layer* or *planar chromatography*, tlc) or a sheet of cellulose material (*paper chromatography*, pc) or the stationary phase may be packed into a glass or metal column (*column chromatography*, cc). If the stationary phase is a liquid it must be immobilised in some way (otherwise it would not remain stationary for very long!) This can be achieved by coating or chemically bonding the liquid stationary phase to an inert support material which is then spread onto a flat plate or packed into a column. Gas chromatography is carried out only in columns since it would not be possible to restrict the gas to the surface of a flat plate as required in tlc. In liquid chromatography, both thin-layer and column chromatography are commonly used but with the emphasis on column chromatography. 'Classical' liquid chromatography was carried out on relatively wide bore (1–2 cm) columns using gravity flow, but modern *high performance liquid chromatography* (hplc) uses columns which are quite narrow (commonly 4.6 mm internal diameter) through which the mobile phase is pumped at flow rates of 1–5cm^3 min^{-1}. The ability to use narrower bore columns through which the mobile phase flows at high flow rates leads to very efficient separations and relatively short analysis times.

Our definition of chromatography has made no reference to the nature of the sample. If a separative technique is to be widely applicable then there should be little restriction on the type of sample which can be analysed.

∏   Can you think of any restriction that might apply in (*i*) gas chromatography and (*ii*) liquid chromatography?

(*i*) In gas chromatography, as the name implies, the mobile phase is gaseous and therefore the sample must also be gaseous, or at least it must be sufficiently volatile so that it can be vaporised without thermal decomposition. When you heat chemical substances two things may occur. One is that they may decompose into more simple molecules and the other is that they may undergo reaction with other substances. Both of these are usually undesirable from a chromatographic point of view since, although we could still separate and identify the reaction or decomposition products, deciding on the course of the reaction and hence the starting materials would be almost impossible. There are techniques, eg pyrolysis gas chromatography of polymers, where the sample is deliberately decomposed and a 'finger-print' pattern of the products is obtained which can be used to identify an unknown polymer by matching up the 'finger-print' with those of standards.

(*ii*) In liquid chromatography, the only requirement is that the sample is soluble in the mobile phase. If it were not soluble, it could not be transported through the column by the mobile phase. Many biochemical molecules are either too involatile or are decomposed on heating, eg proteins, and cannot be separated by gc, but they can be separated using liquid chromatography with a suitable mobile phase.

1.2. CLASSIFICATION OF CHROMATOGRAPHIC TECHNIQUES

We have already seen that chromatography may be classified into gas and liquid chromatography according to the nature of the mobile phase, and that this classification may be extended by considering whether the stationary phase is a liquid or a solid. This approach is rather too general, and further systems have been suggested.

∏ Can you suggest any features of chromatography which could be used as a basis for classification?

We have already seen one basis for a classification according to whether the stationary phase is used in the form of a column or a flat sheet. This is a classification according to technique.

1.2.1. Classification by Technique

A simplified diagram of a chromatographic column and a thin-layer plate is shown in Fig. 1.2a.

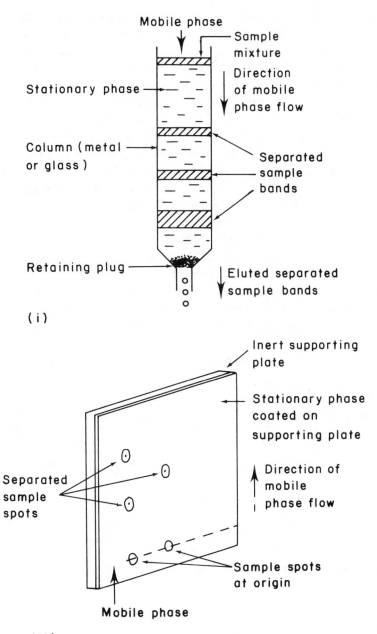

Fig. 1.2a.   *(i) column chromatography and (ii) thin-layer chromatography*

Column chromatography may be further sub-divided according to the nature and dimensions of the column. If the stationary phase is coated or bonded onto an inert solid support and then packed into the column then the column is referred to as a *packed column*. However, another way to hold the stationary phase in the column is to spread it as a thin film onto the internal wall of a length of narrow bore tubing. These columns are called *capillary* or *open tubular columns*. If the walls of the tubing are used without the addition of any material that could be considered as solid support the columns are called *wall-coated open tubular (WCOT) columns*. If the walls are first coated with a layer of fine particulate support material which is then coated with the stationary phase, the columns are called *support-coated open tubular (SCOT)* columns. If the column wall is extended by the addition of a porous layer (eg fused silica or glass powder) the column is called a porous-layer open tubular (PLOT) column. The term *capillary column* usually refers to columns with a very small internal diameter (< 0.5 mm i.d.). *Microbore* columns have internal diameters in the range 0.5-1.0 mm and *conventional* columns have diameters > 1.0 mm.

The various forms of columns are illustrated in Fig. 1.2b.

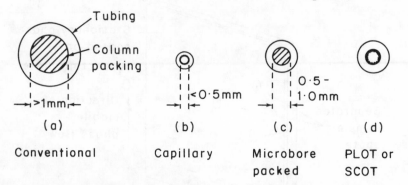

Fig. 1.2b.   *Different forms of column used in chromatography*

1.2.2.   Classification According to Development Method

In chromatography, the term *development* is used to describe the manner in which the sample is carried through (or *eluted* from) the column by the mobile phase. Although the first of these methods, *elution development* is used in most separations, the other two, *frontal analysis* and *displacement development* are still used in some instances, eg frontal analysis is still used to obtain thermodynamic data from chromatographic measurements, and displacement development is used in some preparative scale separations using column chromatography.

(1) Elution development

In this method, the sample is injected into the mobile phase as a discrete plug or band of material and is swept through the column by a continuous stream of mobile phase which has a lower affinity for the stationary phase than any components in the sample. If the sample components have different rates of migration through the column, they may be separated into separate bands with zones of pure mobile phase between them. (This is the situation represented in Fig. 1.2a.) The sample components are not obtained 'pure' since each sample band is mixed with mobile phase, but this may be removed by a physical separation method.

Π   What physical methods can you think of which you could use to separate

    (*i*)   a gaseous sample in a gaseous mobile phase and

    (*ii*) a liquid sample from a liquid mobile phase?

(*i*)   The first question obviously refers to gas chromatography. In Section 1.1 we saw that the sample in gas chromatography must be a gas or a volatile liquid or solid. The sample and carrier gas could be separated by condensing out the volatile sample component by passing the eluate through a cold trap, since there is usually a large difference in the boiling points of the carrier gas and the sample components. With gaseous samples, the boiling point differences may be small and a very low temperature would be required, but in principle the method would still work. In practice, however, the recovery of vapours from a gas stream is not an easy matter.

(*ii*) In liquid chromatography again there is usually a large difference in boiling points between the sample components and the
mobile phase so that simple evaporation or distillation of the
mobile phase is all that is required.

If the composition of the mobile phase is unchanged during an analysis the technique is called *isocratic* or *simple elution*. This technique is always used in gas chromatography since the only function
of the carrier gas is to transport the sample through the column.
Simple elution can also be used in liquid chromatography for the
separation of relatively simple mixtures which elute fairly rapidly,
but for more complex mixtures or for mixtures in which some components elute only slowly it may be preferable to change the composition of the mobile phase during the course of the analysis. This
may be done at particular times (*stepwise elution*) or in a continuous
manner (*gradient elution*). By changing the composition of the mobile phase, its 'strength' or 'eluting power' can be changed, and this
changes the rate of migration of components through the column.
(The reason for this will become clear when we look at Section 1.4.)

The time taken for a component to be eluted from the column (the
elution or *retention time* t_R) under a given set of conditions (eg
stationary phase, temperature, flow rate etc) is characteristic of that
component (though not necessarily uniquely characteristic) and can
be used in the identification of an unknown component in a mixture.

(2) *Frontal Analysis*

In frontal analysis the sample is swept continuously onto the column by the mobile phase. All the sample components will be held
on the column until the column becomes saturated with respect to
a particular component. That component is then eluted from the
column. The component with the smallest affinity for the column
will be eluted first, initially in a pure form, then as a mixture with
the next component to be eluted, and so on until all the components have been eluted. The time taken for a component to 'break
through' the column is the retention time t_R, and it can also be used
in a qualitative way to identify the components in a mixture.

(3) *Displacement Development*

In this method, the sample is again placed on the top of the column as a discrete plug, but unlike elution development the mobile phase is a substance which has a higher affinity for the stationary phase than any component in the mixture. The mobile phase or *displacer* thus pushes the component bands off the column and they move down the column and are separated according to their affinity for the column, each component acting as a displacer for the component ahead of it. Again the break-through times may be used to help identify the components of a mixture. Although each component is 'pure' as it is eluted from the column there are zones of overlap between the pure components containing a mixture of the two adjacent zones.

SAQ 1.2a

Complete the following diagrams showing how the components A, B and C of a mixture will appear as they pass down the chromatographic column in (*i*) elution development (*ii*) frontal analysis and (*iii*) displacement development. The affinity of the components for the column is in the order A < B < C.

⟶

SAQ 1.2a
(cont.)

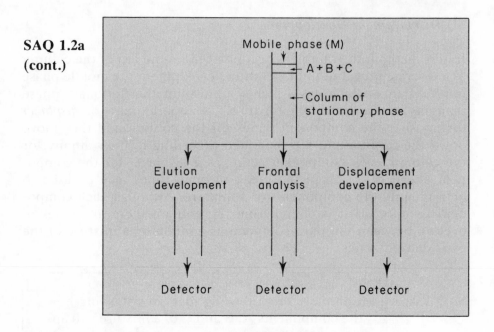

SAQ 1.2b   Following on from SAQ 1.2a, if a detector whose response is a function of component concentration is used to detect the eluent, complete the following diagrams to show how the signal will vary as the components A, B and C are eluted.

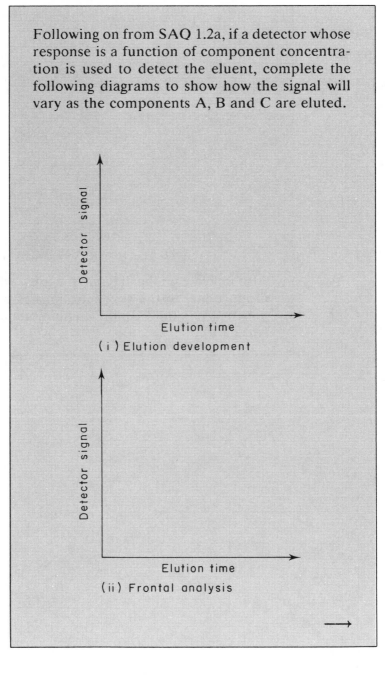

SAQ 1.2b
(cont.)

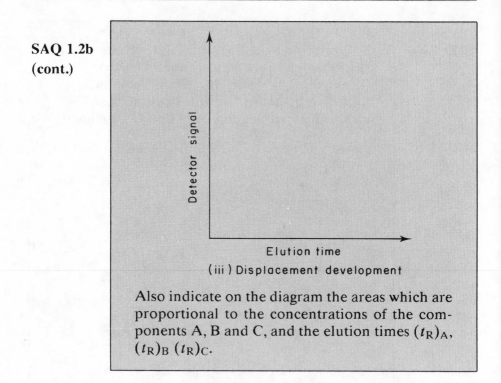

(iii) Displacement development

Also indicate on the diagram the areas which are proportional to the concentrations of the components A, B and C, and the elution times $(t_R)_A$, $(t_R)_B$ $(t_R)_C$.

1.2.3. Classification According to the Mode of Separation

As a sample moves through the system, the components of the sample interact with the stationary phase. There are several different types of interaction which may occur during the separation and these are often referred to as *modes* of separation, eg adsorption, partition, ion-exchange etc. These terms will be discussed in Section 1.5. Suffice to recognise at the moment that the different modes of separation themselves form a basis for classification. Since these terms refer to the basic mechanisms, or physico-chemical processes which occur on the column, it is the most fundamental method of classifying chromatography. However, this approach also has its problems since in some instances it is not clear what the predominant mode of separation is, eg if using silica gel thin-layer plates it is obvious that adsorption will be a possible mechanism by which the molecules will interact with the stationary phase. However, silica has a strong affinity for water so that water molecules in the atmosphere would be adsorbed onto the silica gel plate even though the plate may have been 'dried' before use. The water molecules will act as a liquid stationary phase and the sample molecules may partition into the water giving a mixed mechanism of retention (adsorption + partition). Similarly a stationary phase which has been prepared by chemically bonding a liquid to a silica surface will not have the same properties as the bulk liquid, ie the process of chemically bonding the molecule to the silica surface has modified the characteristics of the liquid phase. In spite of these problems, it is convenient to be able to describe a separation as being due to adsorption, partition, ion-exchange etc even though these may not be the only mechanisms operative. These various methods of classifying chromatographic separations are bought together in Fig. 1.2c. The table is not all embracing but this only serves to show how difficult it is to draw up a simple classification for a process which is anything but simple.

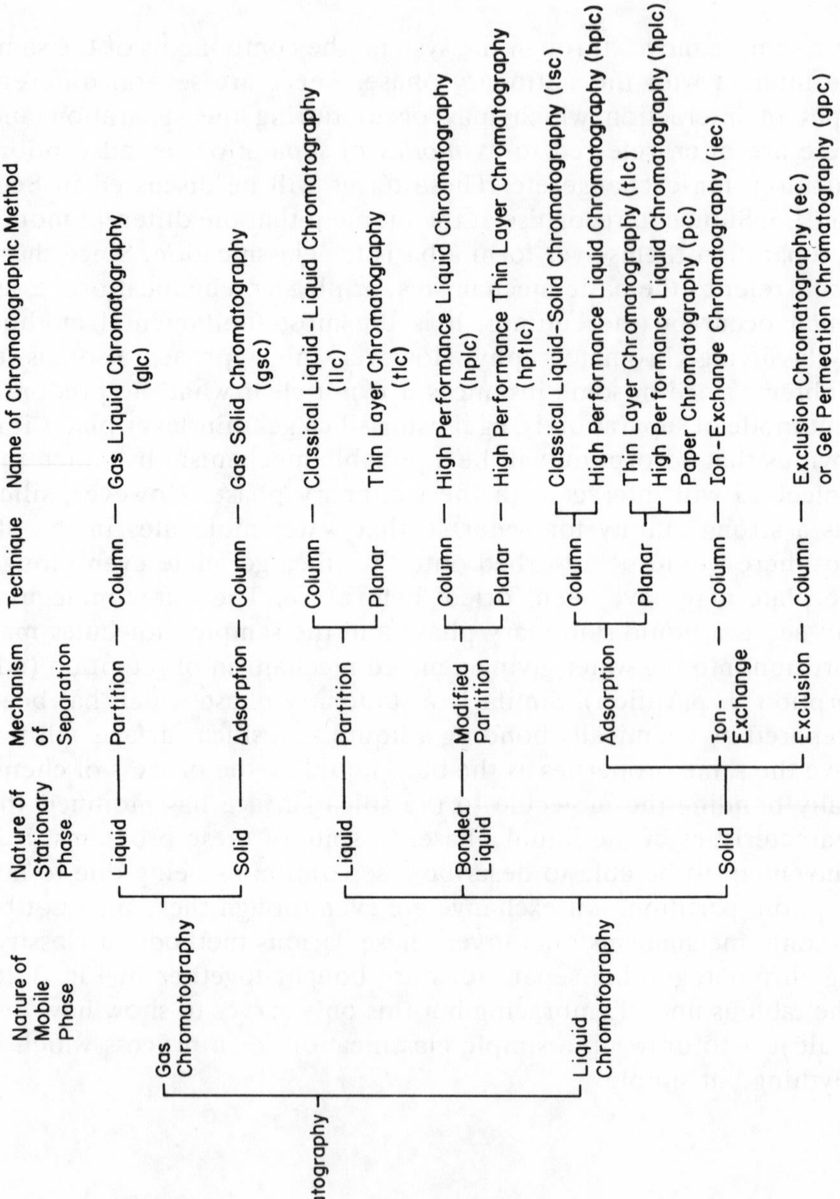

Chromatography

Nature of Mobile Phase	Nature of Stationary Phase	Mechanism of Separation	Technique	Name of Chromatographic Method
Gas Chromatography	Liquid	Partition	Column	Gas Liquid Chromatography (glc)
	Solid	Adsorption	Column	Gas Solid Chromatography (gsc)
Liquid Chromatography	Liquid	Partition	Column	Classical Liquid–Liquid Chromatography (llc)
			Planar	Thin Layer Chromatography (tlc)
	Bonded Liquid	Modified Partition	Column	High Performance Liquid Chromatography (hplc)
			Planar	High Performance Thin Layer Chromatography (hptlc)
	Solid	Adsorption	Column	Classical Liquid–Solid Chromatography (lsc)
			Column	High Performance Liquid Chromatography (hplc)
			Planar	Thin Layer Chromatography (tlc)
			Planar	High Performance Liquid Chromatography (hplc)
				Paper Chromatography (pc)
		Ion-Exchange	Column	Ion-Exchange Chromatography (iec)
		Exclusion	Column	Exclusion Chromatography (ec) or Gel Permeation Chromatography (gpc)

Fig. 1.2c. *Classification of chromatographic systems*

1.3. DISTRIBUTION COEFFICIENT

A common laboratory technique for the purification and separation of chemical compounds is that of *solvent* or *liquid-liquid* extraction. The technique is based on the distribution of a solute between two essentially immiscible liquids one usually being an organic solvent and the other aqueous. At equilibrium, a solute which is soluble in both phases will distribute between the two phases in a fixed proportion according to the distribution or partition coefficient (K_D) where

$$K_D = [A]_o / [A]_{aq}$$

and the square brackets denote concentrations (strictly activities) in the organic and the aqueous phases.

The distribution coefficient is independent of the amount of solute taken and also of the amount of solvent used as it is expressed in terms of concentrations (eg g dm^{-3} or mol dm^{-3}). It is also constant provided that the temperature and pressure remain constant, the solvents are immiscible and do not react with each other and that the solute molecules do not react, associate or dissociate in either solution.

Components can be selectively extracted from aqueous solutions into organic solvents or re-extracted from the organic phase into the aqueous one. The required separation can be achieved by adjustment of the chemical parameters; pH, ionic strength, nature of the solvent and the addition of masking agents.

The completeness of the extraction depends not only on the value of the distribution coefficient but also on the volumes of the phases used. Most organic molecules are an order of magnitude more soluble in an organic solvent than in water ie $K_D \geq 10$, but this can be affected dramatically by the nature of any functional groups in the molecule eg, an alcoholic OH or a carbonyl group will reduce the distribution coefficient by a factor of 5–150 because of the formation of hydrogen-bonds with the aqueous phase. Introduction of a halogen atom increases the distribution coefficient by a factor of between 4 and 40.

The completeness of an extraction as measured by the concentration (C) of solute left in the aqueous phase after one equilibration using volume V of aqueous and volume V_0 of organic phase is given by

$$C = C_0 \frac{1}{1 + K_D (V_0/V)}$$

where C_0 is the initial concentration of solute, so that the amount remaining after a single extraction is clearly dependent on the value of K_D and the volume ratio of the two phases. We shall see later that these same two parameters figure prominently in the equations for chromatographic retention.

∏   Can you suggest a means by which the purity of a sample could be improved using solvent extraction?

Since the purity achieved in a single extraction is clearly limited, an obvious improvement would be to use repeated batch extractions or a continuous extraction method.

In the continuous extraction method, fresh portions of organic phase are brought into contact with the aqueous phase in a multiple partition process with a large number of stages, which are discontinuous and stepwise in nature. This technique, known as *counter-current distribution*, has been combined with liquid chromatography to give *counter-current chromatography* (ccc). It involves a liquid stationary phase through which droplets of the mobile phase pass. Chromatographic separation of solute molecules occurs according to the solute partition coefficients as in column liquid chromatography.

We can look on our chromatographic column or flat layer of stationary phase eg, gc column or thin-layer plate as being analogous to a series of separating funnels containing the two phases, one stationary and the other mobile. We can re-define the distribution or partition coefficient as

$$K_D = C_S / C_m$$

where C_s is the concentration of a component in the stationary phase and C_m is the concentration of the same component in the mobile phase.

SAQ 1.3a

The following results were obtained for the equilibrium concentrations of 2,4-pentanedione after extraction by tetrachloromethane from water.

Extraction	[2,4-pentanedione] in CCl_4/mol dm^{-3}	[2,4-pentanedione] in water/mol dm^{-3}
1	0.727	0.360
2	0.842	0.421
3	1.010	0.515

Calculate the distribution coefficient for 2,4-pentanedione between CCl_4 and water.

SAQ 1.3a

SAQ 1.3b

3 g of iodine is dissolved in a mixture of water and benzene (100 cm^3 of each) On analysis, 1 g of iodine is found in the water layer. If 5 g of iodine were dissolved in a mixture of water and benzene (100 cm^3 of each) at the same temperature, how much iodine would be found in the water layer?

1.4. RETENTION IN CHROMATOGRAPHY

Fig. 1.4a shows what happens when a sample is introduced onto a column in the mobile phase. In practice, the stationary phase will be dispersed throughout the column but for convenience here we shall imagine that it occupies one half of the column and that the mobile phase occupies the other half.

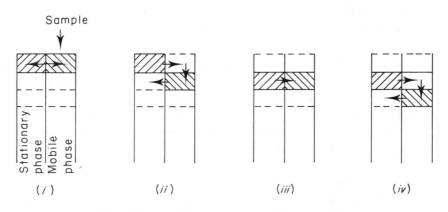

Fig. 1.4a. *Chromatographic retention*

When the sample (carried by the mobile phase) comes into contact with the stationary phase it will distribute between the two phases according to the value of its distribution coefficient (K_D or K). This is equivalent to the separation in our first separating funnel. However, unlike our separating funnel, we do not have to wait to open the tap on the funnel to allow the sample to pass into the second funnel. Neither can we wait for our sample to reach its equilibrium distribution between the two phases before passing it on to the second separation step. This means that if the equilibrium distribution is to be achieved it must occur almost instantaneously. Having been distributed between the two phases, the sample remaining in the mobile phase will be moved further down the column where it will meet fresh stationary phase. The sample will again move from the mobile phase into the stationary phase (this process is called *sorption*) when equilibrium will again be achieved. Because the concentration of the sample in the mobile phase in the first section has been reduced, sample will move from the stationary phase back into the mobile phase (this process is called *desorption*) in order to keep the value of the distribution coefficient constant. This process is illustrated in Fig. 1.4a(*ii*). In a very short time all the sample will have desorbed from the stationary phase in the first section of the column and will be swept into the second section of the column by the mobile phase. Eventually, equilibrium will again be established and the sample will be found completely in the second section of the column (Fig. 1.4a(*iii*)). This process of sorption and desorption will continue (Fig. 1.4a(*iv*)) until the sample reaches the end of the column, or is *eluted* from the column.

Chromatographers do not refer to separating funnels or sections of the column but borrow terminology from the technique of distillation with which chromatography has certain common features. By analogy with distillation, each section of the chromatographic column in which equilibrium is assumed to be achieved is known as a *theoretical plate* (see also Sections 3.4 and 3.5).

We have looked on the chromatographic process as occurring in a series of discrete steps as with the separating funnel. But chromatography is a dynamic process not a series of static ones. However, by making our 'steps' or 'theoretical plates' extremely small the static and dynamic processes become equivalent. The length of column corresponding to a theoretical plate is called the *H*eight *E*quivalent to a *T*heoretical *P*late (or HETP or simply *H*). The smaller the value of *H* the more equilibration steps there are in the column (this is equivalent to using a large number of successive separations in a funnel) and the column is said to possess a high efficiency. In practice, this means that the column should be better at separating the components of a mixture.

Referring again to Fig. 1.4a, if the arrows indicate the direction of movement of the solute particles, we notice that there is no movement directly down the column in the stationary phase. The particles move only across the boundary between the stationary phase and the mobile phase, and down the column during the time they spend in the mobile phase. In the mobile phase, all particles will move at the same speed ($\bar{u}_m$), where $\bar{u}_m$ = linear velocity of the mobile phase, so that the time spent in the mobile phase is the same for all particles. This time is called the *hold-up time* or *column dead time* (t_m).

Differential rates of migration are therefore due to the time (t_s) that different species spend in the stationary phase. A component (A) with a low distribution coefficient, K, will have fewer of its particles in the stationary phase at any given time compared with another component (B) with a higher distribution coefficient and will therefore be retarded less and will migrate through the column more rapidly. Thus, the rate of migration of a component, R_m, is inversely proportional to its distribution coefficient, ie rate of migration

$$R_m \propto \frac{1}{K}$$

and the components of a mixture will be separated only if they have different distribution coefficients between the stationary and the mobile phase.

The overall retention time of a component (t_R) is given by the sum of the times spent in the mobile and stationary phases such that

$$t_R = t_m + t_s$$

(t_s is usually called the *adjusted retention* time, t'_R.

In gas chromatography the mobile phase (usually nitrogen, helium or hydrogen) is said to be non-interactive, ie it does not affect the distribution of the sample molecules between phases which depends only on the molecular interactions between the sample and the stationary phase. The only function of the mobile phase is to transport the solute molecules through the column. In liquid chromatography the mobile phase is described as being interactive and it plays an important part in the distribution of the sample components and hence in the separation. By changing the chemical composition of the mobile phase, the distributions of components between the stationary and mobile phases and hence their rates of migration may be altered considerably. Thus in liquid chromatography, the chromatographer has additional control (or selectivity) over the separation achieved.

1.5. SORPTION MECHANISMS

One reason for the success of chromatography as an analytical technique lies in the several different mechanisms or modes which can be exploited and which make it possible to achieve separations of such diverse types as apolar, polar, ionic, ionisable and polymeric species. It is easier to understand a separation if only a single mode is in operation, it being more difficult to predict retention behaviour if more than one mode of chromatography is exploited, either by accident or design.

There are several modes of separation:

(*i*)   adsorption

(*ii*)   partition

(*iii*)   bonded phase

(*iv*)   ion-exchange

(*v*)   ion-pairing

(*vi*)   exclusion

(*vii*)   affinity

The first five modes depend on interactions between the sample molecules or ions and the stationary and the mobile phases. Exclusion chromatography separates molecules which differ in size and shape, whilst affinity chromatography makes use of very specific biochemical interactions between, for example, a ligand, eg $-NH(CH_2)_6.NH.C.OR$, attached to a matrix and say a nucleic acid or a protein molecule.

∏   Why do you think it necessary to exploit so many different modes of chromatography?

     As we shall see later (Section 2.3.1) a chromatographic separation depends on interactions between the sample components and the stationary phase (and in liquid chromatography, the mobile phase as well). With very diverse sample types, it is not always possible to make use of a particular type of interaction, eg ionic sample species will not interact with an apolar stationary phase so that we may need to use the ion-exchange or ion-pairing modes in this case. It is the versatility of liquid chromatography in particular which makes it such a powerful separative technique.

1.5.1. Adsorption

Adsorption in the form of liquid-solid chromatography (lsc) is the oldest technique and was used by M. S. Tswett in 1906 to separate plant pigments. In the form of thin-layer chromatography (tlc) it was introduced by Kirchner in the early 1950s and as high performance liquid chromatography (hplc) it was first used by several workers in 1969.

In the adsorption process, solute and solvent molecules compete for 'sites' on the adsorbent; to be adsorbed, the solute molecule must first displace a solvent molecule. If the adsorbent is a polar material (eg silica or alumina) non-polar molecules (eg hydrocarbons) will have little affinity for the surface and will not be retained. Molecules with polar functional groups or those capable of H-bonding will have a strong affinity for the adsorbent surface and will be strongly retained. Polarisable molecules, eg aromatic compounds and others with a conjugated π-electron system, will undergo dipole/induced dipole interactions (see Section 2.3.1) with the adsorbent surface and will be retained, the degree of retention depending on the polarisability of the functional group or molecule. With apolar adsorbents (eg graphitised carbon) the dominant intermolecular forces are the London (dispersion) forces and polar and polarisable molecules will be less strongly retained.

The adsorbent surface consists of discrete adsorption sites. In the case of silica, these are hydroxyl ($-OH$) groups, the grouping $-Si-OH$ is known as a silanol group. Two silanol groups may be dehydrated (ie they lose a molecule of water) to give a siloxane group, ie

$$2 \, (\geqslant Si-OH) \longrightarrow \geqslant Si-O-Si \leqslant + H_2O$$

and the spacing of the —OH groups is determined by the Si—Si bond distance. This gives rise to the possibility of selective adsorption when the stereochemistry of the surface and solute molecules are suitably matched. With geometric isomers, eg 1,2- 1,3- and 1,4-dibromobenzene, the 1,4-isomer is able to interact with two surface —OH groups and the 1,2-isomer with only one —OH group. The 1,3-isomer is intermediate in interaction and therefore the order of retention is 1,4- > 1,3- > 1,2-. Adsorption chromatography is therefore suited to the separation of geometric isomers.

Since the forces between the solute and adsorbent molecules depend on polarity effects, it follows that adsorption chromatography is better suited to the separation of a mixture into molecular types eg, alcohols, ketones, esters etc., than for the separation of members of a homologous series where the affect of an additional —CH$_2$-group in a hydrocarbon chain produces insufficient differences in the strengths of the intermolecular forces to allow a separation. The order of elution of polar molecules on a polar stationary phase is predictably in the order of the polarity of the solute functional groups, ie

$$-CH=CH-<-OCH_3 <-CO_2R <-C=0$$
$$<-CHO <-SH <-NH_2 <-OH <-CO_2H$$

The order is reversed for an apolar adsorbent (eg carbon). A practical problem with liquid-solid chromatography is the effect that water has on the activity of polar adsorbents. The water is adsorbed onto the strongest adsorption sites leaving a more even distribution of weaker sites that retain the sample. This leads to a decrease in retention times for other molecules and the adsorbent is said to be *deactivated*. For strongly polar solute molecules, the consequent reduction in retention may be desirable but the problem arises in controlling the amount of water present. This may vary because of:

(*i*)   variations in the amount of water present in supposedly 'dry' solvents.

(*ii*)   changes in the water content of the mobile phase with time depending on the relative humidity of the atmosphere and the water content of the solvent.

(*iii*)   removal of water from the walls of containing vessels etc by the solvent.

By adding water to the mobile phase to give a solution which is, say, 30% saturated with water, the effect of further changes in water content is markedly reduced. The beneficial effects of added water are:

(*i*)   less variation in sample retention times from run to run.

(*ii*)   higher sample loadings possible resulting from a more linear 'water modified' isotherm (see Section 3.3).

(*iii*)   more symmetrical peaks and more efficient columns.

(*iv*)   reduction in batch-to-batch variations of the adsorbent.

(*v*)   reduced catalytic activity of the adsorbent.

Other factors which affect the retention of sample molecules are the surface area and average pore diameter of the adsorbent. To a first approximation, the retention is proportional to the surface area, (typically 200–250 $m^2 g^{-1}$ for a spherically shaped silica particle and of the order of 400 $m^2 g^{-1}$ for an irregular shaped particle). The pore diameter has little effect on the retention unless the diameter is so reduced that larger molecules are excluded from the pores. This reduces the effective surface area of the adsorbent and a decreased retention is observed. In this case, the mechanism for the retention would be a mixed one of adsorption and exclusion. Typical pore diameters are in the range 5–10nm (50–100 Å).

1.5.2.   Partition

In partition chromatography, the solid adsorbent is replaced by a liquid stationary phase. In gas-liquid chromatography, the liquid phase is coated onto an inert solid support (usually a silaceous material). Whilst partition chromatography is the most widely used mode in gas chromatography, in high performance liquid chromatography difficulties arise from the tendency for the liquid stationary phase to be stripped from the column. This is due to the solubility of the stationary phase in the mobile phase and to the sheer forces which act on the stationary phase due to the movement of the mobile phase.

∏   Why can we use a liquid stationary phase in gas chromatography?

Remember that in gas chromatography by definition the mobile phase is a gas. Therefore the liquid stationary phase will not show any solubility in the mobile phase nor will the moving gas flow cause sufficient sheer force to remove the liquid.

A liquid stationary phase can be used in some forms of liquid chromatography, eg in thin-layer chromatography, but even that is rare today with the introduction of bonded stationary phases (see Section 1.5.3).

Partition chromatography utilises the ability of a solute to distribute itself between two phases according to its distribution or partition coefficient as in liquid-liquid extraction. Solutes with a high solubility in the stationary phase will have a high partition coefficient since

$$K = \frac{\text{concentration of sample in the stationary phase}}{\text{concentration of sample in the mobile phase}}$$

$$= \frac{C_s}{C_m}$$

∏   Will a solute with a high distribution coefficient elute slowly or rapidly from the column?

In Section 1.4 we said that a high distribution coefficient means that the solute molecules spend most of their time in the stationary phase and will elute slowly from the column since movement through the column occurs only when the molecules are in the mobile phase.

Dispersion, dipole/dipole, dipole/induced dipole and H—bonding forces (see Section 2.3.1) will all play their part in determining the solubility of a solute, and partition chromatography is suitable for the separation of a wide range of molecular types and has the advantage over adsorption chromatography of being able to separate the different members of a homologous series. Furthermore, it does not suffer from problems with water content of the mobile phase, nor from the poor batch-to-batch reproducibility associated with the production of adsorbents. The range of liquids suitable for use as a stationary phase is very large, the only requirement in glc being that the stationary phase is involatile at the column temperature. In llc the range of liquid phases used is relatively small mainly because selectivity can be introduced into the separation by changing the composition of the mobile phase rather than by changing the nature of the stationary phase.

1.5.3. Bonded-phase Chromatography

To overcome the problems of using a liquid as stationary phase, methods of chemically bonding the stationary phase to the silica support to produce *bonded phases* have been developed. These methods which produce hydrolytically stable chemical bonds, have been extended to the production of bonded-phase silica capillary columns for use in gas chromatography, not this time, to prevent the mobile phase from leaching the stationary phase from the column but to prevent the solvent in which the sample has been dissolved from leaching out the stationary phase and also to increase the thermal stability of the phase. Bonded phases are prepared by reacting the surface silanol groups with a chlorosilane. Both monofunctional and bifunctional silanes can be used.

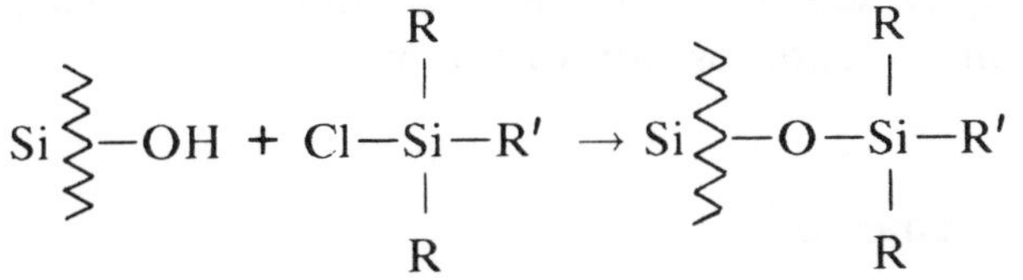

$$
\mathrm{Si}\!\!\gtrless\!\!-\mathrm{OH} \;+\; \mathrm{Cl}-\underset{\underset{\textstyle R}{|}}{\overset{\overset{\textstyle R}{|}}{\mathrm{Si}}}-R' \;\rightarrow\; \mathrm{Si}\!\!\gtrless\!\!-\mathrm{O}-\underset{\underset{\textstyle R}{|}}{\overset{\overset{\textstyle R}{|}}{\mathrm{Si}}}-R'
$$

Fig. 1.5a. *Reaction of a silanol group with a monofunctional silane*

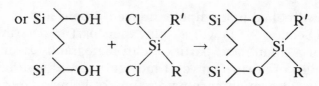

Fig. 1.5b.   *Reaction of silanol groups with a bifunctional silane*

The groups R are usually methyl (CH$_3$) groups. The nature of R′ can be varied to give both apolar and polar stationary phases. Commonly, R′ is a hydrocarbon chain (C$_6$, C$_8$ or C$_{18}$) which gives rise to an apolar phase, but polarity can be introduced into the molecule by substituting the terminal—CH$_3$ group in the hydrocarbon chain by for example a nitrile group ($-C\equiv N$) or by an amino group ($-NH_2$). If the stationary phase is more polar than the mobile phase the technique is referred to as *normal phase chromatography* and if the stationary phase is less polar than the mobile phase the technique is called *reverse phase chromatography*.

The mechanism of bonded-phase chromatography (bpc) is complex. The bonded liquid, whilst retaining many of the characteristics of the bulk liquid, shows sorptive properties which are perhaps best described as a modified partition. In reverse-phase chromatography, retention occurs by non-specific hydrophobic interactions of the solute with the stationary phase and the very wide application of reverse-phase chromatography is due to the fact that virtually all organic molecules have hydrophobic regions in the structure that are capable of interacting with the stationary phase.

The range of bonded-phase packings commercially available is rather limited, the main reason being that, as in liquid partition chromatography, selectivity in the separation is more easily achieved by changes in the composition of the mobile phase.

1.5.4.   Ion-exchange

Ion-exchange has been used in industry (eg in water softening) for some time and it is not surprising that the technique should be applied to chromatographic separations. Ion-exchange involves sub-stitution of one ionic species for another. The stationary phase is a rigid matrix (M) the surface of which carries positive or negative charges to give ion-exchange sites (R^+ or R^-). Counter-ions of op-posite charge (Y^- or Y^+) are associated with each site in the matrix and these can exchange with similarly charged ions in the mobile phase to be held on the exchange site. Sample ions (X^- or X^+) may thus exchange with these counter ions (Y^- or Y^+):

$$MR^+Y^- + X^- \rightleftharpoons MR^+X^- + Y^-$$

or $$MR^-Y^+ + X^+ \rightleftharpoons MR^-X^+ + Y^+$$

If the process involves the exchange of negatively charged ions it is known as *anion-exchange*. The complimentary process is known as *cation-exchange*.

∏   What type of sample could you separate using ion-exchange chromatography?

As explained, the mechanism involves the exchange of sample ions with the counter-ions, therefore the sample must either be an ion or a molecule which is capable of giving an ionic species at a certain pH, ie an ionisable molecule, such as an organic acid (HA) or base (BH).

$$HA \rightleftharpoons H^+ + A^-$$

or $$B + H^+ \rightleftharpoons BH^+$$

∏   In adsorption chromatography, the retention of the sample depends on the affinity of the sample for the adsorbent sur-face. What will determine the retention in ion-exchange?

The retention will depend on the strength of interactions between the sample ions and the exchange sites. Ions that react only weakly with the exchange sites are poorly retained and will be eluted rapidly whilst ions with strong interactions will elute slowly. The degree of interaction (and hence retention) will depend on the nature of the functional groups on the resin and also on the separating ions. Cation-exchangers contain acidic groups which may be described as weak (eg $-COO^-$) or strong (eg $-SO_3^-$) whilst anion exchangers contain basic groups again weak (eg $-NH_2$) or strong (eg $-NR_3^+$). The degree of interaction (and hence retention) shows no exact correlation with any simple property of the ions, though both the hydrated ionic radius (which limits the coulombic interaction) and the polarisability of the ions which determines the van der Waals' attractions are important. Thus, for a series of monovalent ions (eg the alkali metals) the order of elution would be $Li^+ < Na^+ < K^+$. For ions of different net charge, the relation between charge and elution order is more complex, but in general we can say that the greater the charge on the ion the more strongly it is retained on the column.

The overall ionic strength, pH and the presence of complexing agents in the mobile phase may have a marked effect on the separation process, eg Cl^- ions may complex with heavy metal cations to give a complex anion which can then be separated by anion-exchange, organic acids may be separated according to their strengths on a strong anion-exchange column, with the weakest acids eluting first, either by displacement development with a stronger acid, or, since acid dissociation depends on the pH, by using a gradient elution with buffer solutions of decreasing pH.

Ligand-exchange is a related technique whereby ligands are separated by virtue of their complexing strength for a metal ion sorbed on the exchange resin, the counter-ion again being displaced.

The matrix in ion-exchange is either a rigid silica bead to which the ion-exchange groups have been chemically bonded or it is a polymer usually based on a copolymer of styrene and divinylbenzene (Fig. 1.5c).

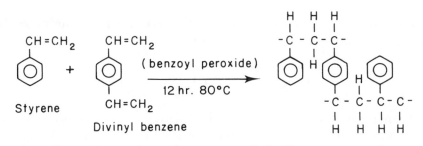

Fig. 1.5c. *Formation of a styrene-divinylbenzene copolymer*

The functional groups are then added into the polymer matrix to give for example:

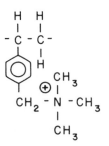

The cross-linking of the polymer provides mechanical rigidity but such resins are susceptible to high pressures and also may swell in the presence of the mobile phase. Ion-exchange chromatography is restricted to liquid chromatography.

1.5.5. Ion-pair Chromatography

Ion-pair chromatography (ipc), sometimes called paired-ion chromatography (pic), can also be used for the separation of both ionic or ionisable molecules, but it has the advantage over ion-exchange of using the same bonded stationary phases that are used in bpc. In its usual form the technique is used with a hydrocarbon bonded stationary phase in the reverse-phase mode. Under these conditions, an ionic or ionisable sample may have insufficient lipophilic character to be retained, ie it is too polar to interact with the apolar hydrocarbon phase. To increase retention, the character of the sample molecule must be altered so as to render it less polar. One approach for an ionisable molecule is to adjust the pH of the mobile phase to suppress the ionisation, ie for the equilibrium

$$R.COOH \rightleftharpoons R.COO^- + H^+$$

addition of acid (ie H^+ ions) will suppress the ionisation so that the equilibrium lies to the left-hand side of the equation. The same result may be achieved by adding an ion-pairing agent to the mobile phase. The ion-pairing agent is a bulky organic molecule having the opposite charge to that of the ion(s) to be analysed. A neutral ion-pair will be formed with the ionic analyte thus

$$
\begin{array}{lll}
A^+ & + \quad B^- & \rightleftharpoons \quad AB \\
\text{ion} & \text{ion-pair} & \text{neutral} \\
\text{sample} & \text{reagent} & \text{ion-pair}
\end{array}
$$

Because the ion-pair reagent contains bulky organic (ie hydrocarbon) substituents, the ion-pair which is formed is hydrophobic in character (ie it is lipophilic), it therefore interacts with the hydrocarbon stationary phase and is retained.

An example of a separation using this technique is the analysis of inorganic anions eg Cl^-, Br^-, SO_4^{2-} etc by forming a neutral ion-pair with the tetra-*n*-butylammonium ion. The formation of the ion-pair can be written

$$(C_4H_9)_4N^+ + Cl^- \rightleftharpoons (C_4H_9)_4 N^+ Cl^-$$

This mechanism of ion-pairing is sometimes referred to as the partition ion-pair mechanism.

An alternative explanation of the process is that the ion-pairing reagent, because of its organic substituents, will bond to the hydrocarbon stationary phase producing a charged site at which ion-exchange can take place as described previously. It is probable that neither mechanism explains all cases in which the technique is used.

1.5.6. Affinity Chromatography

Affinity chromatography is a relatively recent development which makes use of the unique biological specificity of the interaction between a protein molecule and a suitable ligand. The ligand, which exhibits a specific binding affinity for the protein, is covalently bonded to a gel matrix to form the stationary phase. On adding the sample molecules in a suitably buffered mobile phase, those components (eg proteins) with a specific affinity for the ligand become bound and are retained. Unbound material is eluted from the column in the mobile phase. The composition or pH of the mobile phase is then altered to weaken the specific interaction between the protein molecules and the ligands so that the protein is released from the stationary phase matrix and is eluted from the column.

The gel-ligand bond must be stable under the experimental conditions and the ligand-sample bonding reaction must be specific but reversible. In some instances, the gel matrix would interfere with the specific adsorption properties of the ligand. In this case the gel matrix and the ligand must be separated by the inclusion of a 'spacer group'. The affinity process can then be represented as in Fig. 1.5d.

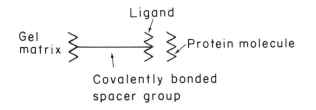

Fig. 1.5d. *Affinity chromatography*

The specific interaction between the ligand and the protein molecule is often referred to as a 'key and lock' mechanism depending as it does on the correct spatial configuration of the substituent groups.

1.5.7. Exclusion Chromatography

Exclusion chromatography differs from other modes of chromatography in that neither specific nor non-specific interactions between the sample molecules and the stationary phase are involved. In fact, attempts are made to eliminate all possible sorption mechanisms from the system.

Exclusion chromatography is carried out on a porous stationary phase which may be based on silica or on a polymer gel. Large molecules having a molecular diameter greater than that of the largest pores are excluded from the pore structure of the gel and pass straight through the column. Very small molecules (eg the solvent molecules) can penetrate into even the smallest pores and will therefore take the longest time to elute from the column.

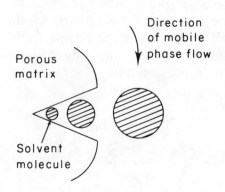

Fig. 1.5e.   *Exclusion mechanism*

Molecules of intermediate size may be able to diffuse into some of the pores and will be eluted at an intermediate time. This process is illustrated in Fig. 1.5e. The large molecules will be eluted first and the smallest ones last. Fig. 1.5f is a calibration curve for an exclusion separation.

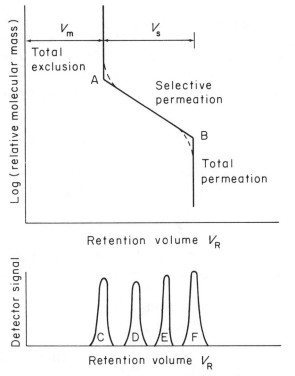

Fig. 1.5f. *Calibration curve for exclusion chromatography*

The point 'A' represents the exclusion limit of the packing. Molecules with a higher relative molecular mass (M_r) than the value given by 'A' will be too large to enter the pores of the gel and are totally excluded. All such molecules in a mixture will be eluted as a single peak (C) and will not be separated from one another.

The M_r given by the point 'B' represents the value below which all the molecules are small enough to permeate the gel totally. All molecules in a mixture with this relative molecular mass or lower will be eluted as a single peak (F) and again will not be separated from one another. Compounds with M_r values between these two extremes will permeate into the pores according to their size and shape and can be separated. Peaks D and E will not necessarily contain a single molecular species but may consist of all molecules within a given M_r range. As a general rule, it is found that a M_r difference between compounds of at least 10% is required before any separation can be achieved. Thus, a peak eluting with a retention volume corresponding to an M_r of, say, 2000 will contain all molecules with values in the range 1900 to 2100. The technique is suitable, therefore, for the separation of polymer species into a series of fractions rather than to the separation of individual polymer molecules.

SAQ 1.5a

Which mode(s) of chromatography would you use to separate the following mixtures?

(i)   inorganic gases, eg CO_2, CO, O_2 and SO_2

(ii)   polyaromatic hydrocarbons, eg anthracene, phenanthrene, pyrene and chrysene

(iii)   fatty acids from an oil, eg palmitic and oleic

(iv)   organo-chlorine pesticides, eg 2,4-D and 2,4,5-T

(v)   a mixture of Cl^-, SO_4^{2-} and CO_3^{2-} ions

(vi)   a mixture of Ca^{2+}, Sr^{2+} and Ba^{2+} ions

(vii)   saturated and unsaturated aliphatic hydrocarbons

($viii$) napthalene sulphonic acids

(ix)   a mixture of steroids

(x)   oligomers of an epoxy resin

(xi)   a mixture of amino acids

(xii)   a mixture of 1,2- 1,3- and 1,4-dihydroxybenzenes

SAQ 1.5a

Learning Objectives

You should now be able to:

● define the term 'chromatography';

● classify chromatographic techniques;

● define the distribution coefficient;

● define the terms 'retention' and 'sorption';

● describe the sorption mechanisms which can be exploited to give chromatographic separations.

2. The Characterisation of Separated Components

Overview

In this section we look at the basic equations which describe the retention of a solute and relate the retention parameters of retention time, retention volume and capacity factor, to the thermodynamic solute distribution coefficient. Intermolecular forces, as they affect retention, are discussed qualitatively.

2.1. RETENTION PARAMETERS FOR COLUMN CHROMATOGRAPHY

2.1.1. The Retention Equations

You may find the mathematical treatment in this Section rather difficult. Regard this Section as optional, but it is important to understand the practical significance of Eq. 2.14 and Eq. 2.16. Attempt the exercise following Eq. 2.16 and SAQ 2.1a and make sure you understand the responses.

We have seen that as the result of time spent in the stationary phase, solute species move through the column at a slower rate ($\bar{u}_x$) than the mobile phase $(\bar{u}_m)$. The ratio of these two velocities is the *retardation factor* (R):

$$R = \frac{\text{rate of movement of solute band}}{\text{rate of movement of mobile phase}} = \frac{(\bar{u}_x)}{(\bar{u}_m)} \quad (2.1)$$

Since solute molecules migrate only when they are in the mobile phase, the speed at which a solute moves through the column depends on the number of molecules in the mobile phase at any instant ie the mole fraction of molecules 'x' in the mobile phase, such that

$$\bar{u}_x = \bar{u}_m \cdot R_x \quad (2.2)$$

where R_x is the mole fraction of x in the mobile phase. If the mole fraction in the mobile phase is zero $(R_x = 0)$ there is no migration as $\bar{u}_x$ is also zero. If the mole fraction in the mobile phase is unity $(R_x = 1)$ then $\bar{u}_x = \bar{u}_m$ and the species x move through the column at the same rate as the mobile phase. Thus, R_x can also be defined as the relative rate of movement of sample species x and is identical to the *Retardation Factor* defined in Eq. 2.1. Therefore we can write:

$$R = \frac{\text{amount of solute in the mobile phase}}{\text{amount of solute in the mobile + stationary phases}} \quad (2.3)$$

If n_s and n_m are the number of moles of solute in the stationary phase and mobile phase respectively, we can write:

$$R = \frac{n_m}{n_s + n_m} \quad (2.4)$$

or dividing by n_m, $\quad R = \frac{1}{1 + n_s/n_m} \quad (2.5)$

The ratio n_s/n_m is called the *capacity factor* (k'), which is related to the retardation factor (R) by:

$$R = \frac{1}{1 + k'} \quad (2.6)$$

From Eqs. 2.4 and 2.6 we can write:

$$R = \frac{n_m}{n_s + n_m}$$

$$= \frac{1}{1 + k'} \tag{2.7}$$

and from Eq. 2.2

$$R = \bar{u}_x / \bar{u}_m$$

$$\therefore \quad \frac{\bar{u}_x}{\bar{u}_m} = \frac{1}{k'}$$

or

$$\bar{u}_m = \bar{u}_x (1 + k') \tag{2.8}$$

For a column of length L, the linear velocity $\bar{u}_x$ of solute x is given by L/t_R where t_R is the retention time of solute x (ie the time taken for x to migrate from the point of injection to the detector) and $\bar{u}_m$ is given by L/t_m where t_m is the hold-up time or column dead time.

Thus

$$\bar{u}_x = L/t_R \text{ and } \bar{u}_m = L/t_m$$

Substituting these two values in (2.8) gives:

$$\frac{L}{t_R} (1 + k') = \frac{L}{t_m}$$

which on rearrangement gives

$$t_R = t_m (1 + k') \tag{2.9}$$

If the concentrations are expressed as moles per unit volume and C_s and C_m are the concentrations of solute in the stationary and mobile phases respectively, then

$$n_s = C_s V_s$$

and

$$n_m = C_m V_m$$

where V_s and V_m are the volumes of stationary and mobile phase respectively.

Hence,
$$k' = \frac{n_s}{n_m} = \frac{C_s V_s}{C_m V_m} \qquad (2.10)$$

but we recognise the ratio C_s/C_m as being the distribution coefficient, K, so that we can write:

$$k' = \frac{K V_s}{V_M} \qquad (2.11)$$

The ratio V_s/V_m is called the *phase ratio*.

Substituting for k' in Eq. 2.9 gives

$$t_R = t_m \left(1 + K \frac{V_s}{V_m}\right) \qquad (2.12)$$

If we measure the retention in volume units (cm^3) rather than in time units (s), the *retention volume* (V_R) is the volume of mobile phase required to elute the solute from the column and is given by:

$$V_R = F \times t_R$$

where F is the volume flow rate of the mobile phase.

Similarly, the total volume (or dead volume) V_m of mobile phase in the column is given by

$$V_m = F \times t_m$$

Substituting for t_R and t_m in Eq. 2.12 gives

$$V_R = V_m \left(1 + K \frac{V_s}{V_s}\right) \qquad (2.13)$$

or
$$\boxed{V_R = V_m + K V_s} \qquad (2.14)$$

This is a fundamental equation in chromatography. It relates the experimental parameters V_R, V_s and V_m to the thermodynamic distribution coefficient K.

Strictly speaking, this equation refers to what is known as 'linear ideal chromatography'. The term 'linear' refers to the assumption that K is a constant and is independent of the solute concentration in the mobile phase, and the term 'ideal' refers to the fact that various processes which broaden the chromatographic band as it passes through the column have been ignored. These topics will be discussed in Section 3.4.

In adsorption chromatography, the volume of the stationary phase, V_s, is not the required parameter since adsorption is not a bulk but a surface phenomenon. Strictly, we should write Eq. 2.14 in the case of adsorption chromatography as

$$V_R = V_m + K_A A_s \qquad (2.15)$$

where K_A and A_s are the adsorption coefficient and adsorbent surface area respectively. With bonded stationary phases in liquid chromatography, it is very difficult to assign a volume to the stationary phase because the orientation of the bonded group on the surface may change with the composition of the mobile phase.

Substituting the values of t_m from $\bar{u}_m = L/t_m$ in Eq. 2.9 and Eq. 2.12 gives:

$$t_R = \frac{L}{\bar{u}_m}(1 + k') = \frac{L}{\bar{u}_m}(+ K\frac{V_s}{V_m}) \qquad (2.16)$$

∏   Eq. 2.16 relates the retention time of a retained compound t_R to important column parameters. What is the effect on t_R of:

(*i*)   an increase by a factor of two in the column length

(*ii*)   a doubling of the mobile phase flow rate

(*iii*) an increase in K

(*iv*) an increase in V_s

(*v*)   an increase in V_m

(*i*)   if we double the column length, Eq. 2.16 shows that the retention time will be doubled since t_R is directly proportional to L.

(*ii*) if the mobile phase flow rate is doubled, the retention time t_R will be halved since

$$t_R \propto \frac{1}{\bar{u}_m}$$

If both (*i*) and (*ii*) are applied simultaneously, the retention time would not be changed.

(*iii*) an increase in K will increase t_R since we saw that the *rate* of movement through the column is inversely proportional to K.

(*iv*) an increase in V_s will also increase t_R. Variations in V_s are sometimes used to control retention in gc. The amount of liquid stationary phase is usually given as a weight per cent on the support material, 10% being a common value. The analysis of isomeric C_3 and C_4 aliphatic hydrocarbons is difficult because the high volatility of these compounds gives short retention times and the column does not have time to produce a separation. Increasing the volume of the stationary phase to say 25% will increase the retention, and separation can be achieved. Similarly, the analysis of steroids even as the trimethylsilyl derivatives leads to long retention times. These can be reduced by using a more lightly loaded column (say 3%).

(*v*)   an increase in V_m will lead to a decrease in t_R. However, this would also have a disastrous affect on the efficiency of the column. Taken to its limit, if V_m was increased until it equalled the total volume of the column there would be no room for stationary phase and no chromatography!

SAQ 2.1a

> The retention times of peaks A, B and C, where A is an unretained component, are 0.84 min, 10.60 min and 11.08 min respectively. If the volume of stationary phase is 12.3 cm^3 and the mobile phase flow rate is 20.95 cm^3 min^{-1} calculate:
>
> (*i*) The retardation factors for B and C and
>
> (*ii*) the distribution coefficients of B and C.

2.1.2. Comparison of Retention Parameters t_R, V_R and k'

All three retention parameters may be measured, the simplest measurement being that of the retention time, t_R. This is measured from the point of injection to the maximum of the chromatographic peak as shown in Fig. 2.1a. For very sharp peaks the maximum is easily determined but for broad peaks it helps to draw in the tangents to the peak and take the point of intersection as shown.

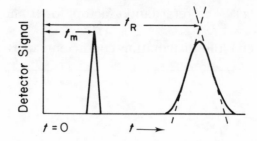

Fig. 2.1a.   *Measurement of retention times*

It is common to measure the distance on the chart recorder and then convert to units of time by multiplying by the chart speed.

It is sometimes preferable to use the retention volume V_R where $V_R = F \times t_R$ and F is the volume flow rate. The use of V_R makes it easier to compare results obtained under different flow conditions. However, the measurement of the volume flow rate whilst simple to do is not always convenient (eg in gc) so that V_R is more suitable for permanent column calibrations. V_R should also be used when flow rate control is not very accurate, since it is independent of F, whereas t_R is dependent on F.

The most commonly used retention parameter in high performance liquid chromatography (hplc) is the capacity factor, k'. Rearrangement of Eq. 2.9 gives.

$$k' = \frac{t_R - t_m}{t_m} \qquad (2.17)$$

where t_m is the retention time of an unretained peak. (ie a solute with $K = 0$, see Fig. 2.1a.) Because k' is a relative value it is less dependent on random variations in the column conditions, eg flow rate and temperature. We shall also see later (Section 3.7.3) that k' occurs in another important equation in chromatography that deals with the resolution of two peaks.

The measurement of t_m requires a solute that has a distribution coefficient $K = 0$. Since the value of K is dependent on temperature a change in column temperature, will produce a change in t_R unless $K = 0$. A temperature change of about 30 °C is usually sufficient to observe this. In gas chromatography, the retention time of methane which is essentially independent of temperature is often taken as t_m. In liquid chromatography, the centre of the first band or baseline disturbance following sample injection is sometimes taken to give t_m. If there is any doubt about this, a substance which is considered to be a weaker mobile phase could be injected, the retention time of this giving t_m; eg if CH_2Cl_2 was the mobile phase with a silica column then n-hexane (less polar) would be considered as a weaker mobile phase and could be used to determine t_m. In reverse phase chromatography with water as the mobile phase, deuterated water (D_2O) could be used to give t_m.

2.1.3. Retention volumes – Definitions

(*i*) *Hold-up volume (V_m)* [sometimes called the column dead volume or void volume]

In Section 1.3 we introduced the hold-up time or column dead time (t_m) as the time that all solutes spend in the mobile phase. The hold-up volume is then given by:

$$V_m = F \times t_m$$

and is a measure of the total volume of space available to the mobile phase in the system, ie the column void plus the injector and detector volumes and the volume of any connecting tubing (eg from the column to the detector). In a well designed system, the extra column void volume is small compared to the column void volume.

The column itself is not filled completely with stationary phase. The fraction of free (non-solid) space within a certain volume element of a porous material is called the *porosity* and is a measure of the space available to the mobile phase (see Section 3.1).

(ii) Retention Volume (uncorrected) V_R

The retention volume as defined in Section 2.1.2 is the volume of mobile phase required to elute the sample from the column and is usually referred to as the retention volume without any qualification though it is properly called the uncorrected retention volume V_R where $V_R = F \times t_R$.

(iii) Adjusted Retention Volume V'_R

In Section 1.4 we saw that the retention time t_R is made up of t_m, the time a solute spends in the mobile phase, and t_s, the time a solute spends in the stationary phase, and that it is the different times that different solute species spend in the stationary phase that leads to the separation. The adjusted retention time is the time that a solute molecule spends in the stationary phase, t_s which is usually given the symbol t'_R ie

$$t'_R = t_R - t_m$$

and the adjusted retention volume V'_R is given by

$$V'_R = F \times t'_R$$

$$= V_R - V_m \tag{2.18}$$

(iv) Corrected Retention Volume V^o_R

This is a term used in gas chromatography, being necessary to take into account the compressibility of the gaseous mobile phase, and is given by

$$V^o_R = j \times V_R \tag{2.19}$$

The term 'j' is called the compressibility factor and is given by

$$j = \frac{3}{2} \left[\frac{(P_i/P_o)^2 - 1}{(P_i/P_o)^3 - 1} \right]$$ (2.20)

where P_i/P_o is the ratio of the inlet to the outlet pressure [For a further discussion of the effect of mobile phase compressibility see Section 2.1.4]

(*v*) *Net Retention Volume* V_N

This is another term used in gas chromatography and is the adjusted retention volume corrected for the compressibility of the gaseous mobile phase ie,

$$V_N = jV'_R$$

$$= j(V_R - V_m)$$ (2.21)

(*vi*) *Specific Retention Volume* V_g

This again is a term used in gas chromatography and allows the comparison of retention data obtained at different temperatures and with different weights of the same stationary phase. It is the net retention volume per gram of stationary phase at 0 °C (273.16 K) and is given by

$$V_g = \frac{V_N \cdot 273.16}{W_s \cdot T_c}$$ (2.22)

where W_s is the weight of stationary phase and T_c is the column temperature.

2.1.4. Mobile Phase Compressibility

Under the conditions used in liquid chromatography the liquid mobile phase is considered to be incompressible. This is not the case with gases.

The mass flow rate of a gas measured in $cm^3\ s^{-1}$ at NTP is constant throughout the column, but because of the compressibility of the gas its linear flow rate (measured in $cm\ s^{-1}$) must change along the column in such a way that it is higher at the outlet than at the inlet. We may represent the column by Fig. 2.1b.

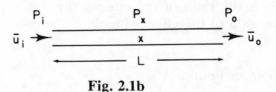

Fig. 2.1b

where $P_i, \bar{u}_i, P_o, \bar{u}_o$ are the pressures and linear velocities at the column inlet and outlet respectively, and P_x is the pressure at a distance x down the column of overall length L.

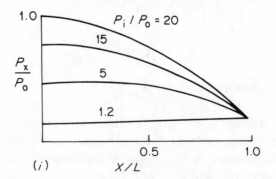

Fig. 2.1c.   *(i) Variation of pressure along a column*

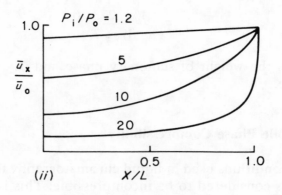

Fig. 2.1c.   *(ii) Variation of flow rate along a column*

From these curves (Fig. 2.1c) we can see that the linear flow rate changes more rapidly along the column when the ratio of P_i/P_o is high, and that the region of most rapid change is at the end of the column.

We shall see later (Section 3.5.4) that a column can operate at its maximum efficiency only at a single value of flow rate and at higher or lower values of this flow rate the efficiency decreases.

∏ Clearly it is important to operate as much of the column as possible in the optimum region. How is this achieved?

Irrespective of the actual value of this optimum flow rate, Fig. 2.1c (*ii*) shows that the ratio P_i/P_o should be as close as possible to unity since the value $P_i/P_o = 1.2$ gives the most constant value of the flow rate throughout the column.

∏ If, in a particular case, it is necessary to have 1/2 bar pressure difference to give the required volumetric flow rate through the column would you choose to have (*i*) $P_i = 1.5$ bar and $P_o = 1.0$ bar or (*ii*) $P_i = 1.0$ bar and $P_o = 0.5$ bar?

The ratio P_i/P_o should be close to unity, thus:

(*i*) $\qquad\qquad P_i/P_o = 1.5/1.0 = 1.5$ bar

or (*ii*) $\qquad\qquad P_i/P_o = 1.0/0.5 = 2.0$ bar

It is therefore better to raise the inlet pressure above atmospheric for the required flow rate than to lower the outlet pressure below atmospheric.

In practical gas chromatography the gaseous mobile phase is available in a convenient pressurised form, and to raise the inlet pressure is the most convenient way to obtain a gas flow. However, in the combination technique of gas chromatography coupled to mass spectrometry (gc-ms), the mass spectrometer which is at the outlet from the chromatography column is operated under high vacuum (ie low pressure conditions) and the ratio P_i/P_o may be quite high resulting in loss of efficiency of the chromatographic column.

SAQ 2.1b   From the given chromatogram calculate for peak B:

(*i*)   the adjusted retention volume

(*ii*) the net retention volume

(*iii*) the specific retention volume

The mobile phase flow rate is 40.0 cm^3 min^{-1}, the chart speed = 0.5 cm min^{-1}, column inlet pressure (P_i) = 1.2 bar column outlet pressure (P_i) = 1.0 bar, column temperature = 100 °C and weight of stationary phase = 1.2 g. Peak A has $K = 0$.

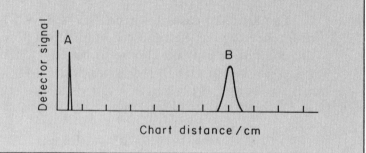

Chart distance / cm

SAQ 2.1b

2.2. RETENTION PARAMETERS FOR PAPER AND THIN-LAYER CHROMATOGRAPHY

The parameter used to characterise the position of a solute in paper (pc) or thin-layer chromatography (tlc) is the retardation factor or R_f value. It is given by the ratio of the distance moved by the solute to the distance moved by the solvent front (Fig. 2.2a).

$$R_f = \frac{\text{distance moved by solute}}{\text{distance moved by solvent front}}$$

the distance being measured to the centre of the spot.

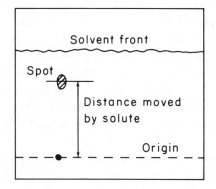

Fig. 2.2a. *Measurement of R_f*

∏   Calculate the R_f value of the solute spot in Figure 2.5

$$R_f = 0.72$$

$$\text{(ie } R_f = \frac{2.1}{2.9} = 0.72)$$

If you got the answer $R_f = 1.38$ you inverted the ratio.

By definition, R_f values are always less than unity. The capacity factor k' of a solute is defined as the ratio of the time spent in the stationary phase to that spent in the mobile phase, so that $k' = t_s/t_m$ and k' and R_f are related by:

$$R_f = \frac{t_m}{t_m + t_s} = \frac{1}{1 + k'} \qquad (2.23)$$

or $$k' = \frac{1 - R_f}{R_f} \qquad (2.24)$$

The term R_f stands for *relative to* (solvent) *front* and is related to the distribution coefficient of the component. Systematic errors in the measurement of R_f values arise from difficulties in locating the exact position of the solvent front and the position of the spot centre. Further errors arise from non-equilibrium between the adsorbent layer, mobile phase and vapour phase. Condensation of the vapour phase or evaporation of the mobile phase in the region of the solvent front will give erroneous values of R_f. Lining the tlc tank with filter-paper is a practical way of attempting to overcome this problem. The filter-paper absorbs the mobile phase and helps to distribute it throughout the tank, saturating the tank with mobile phase vapour. Other factors which affect the reproducibility of R_f values are small changes in the composition of the mobile phase, variations in the properties of the adsorbent, temperature and the size of the tank.

∏   Although the R_f value is not an absolute physical quantity (it
        is related to the distribution coefficient K in a similar manner
        to the retention time t_R) it can be used for identification

purposes with careful control of the operating conditions. Suggest a way in which the reliability of identification could be improved.

Differences in R_f values can be minimised by quoting the R_f value relative to a standard substance run at the same time as the sample and on the same tlc plate. This relative R_f value (or R_{st}) should be constant under any given conditions.

$$R_{st} = \frac{\text{distance moved by solute}}{\text{distance moved by reference standard}}$$

Identification is made more certain by running the sample using different stationary and mobile phases, but the component can be characterised most certainly by removing the spot completely from the plate and then subjecting the separated component to further analysis, eg by infrared, nuclear magnetic resonance or mass spectrometry.

This problem of identification of course relates to all forms of chromatography. Chromatography is a separation technique, and although the retention behaviour of a component under a given set of experimental conditions will give a clue to the general nature of the component, eg high or low relative molecular mass, polarity etc, positive identification by chromatography alone (by comparison with standards) is problematical.

2.3. FACTORS AFFECTING RETENTION

Migration of a solute through a column depends on the equilibrium distribution of the species between the stationary and mobile phases. Retention is therefore controlled by those factors which affect the distribution: the composition of the mobile phase, the nature of the stationary phase and the temperature. Pressure too will, in theory, affect the distribution but at the pressures normally used in gas and liquid chromatography this effect can be ignored.

In gas chromatography, the mobile phase plays no part in the separation and is described as being non-interactive. In liquid chromatography a change in the mobile phase composition is an important way of controlling retention.

2.3.1.　Intermolecular Forces

To understand the factors affecting retention, it is necessary to consider the nature of the forces between molecules. All such forces are electrostatic in origin and are based on Coulombs Law of attraction between unlike and the repulsion between like charges. We can distinguish forces arising from the following interactions:

(*a*)　Coulombic forces between ions

(*b*)　forces arising from interactions between molecules having permanent dipoles (dipole-dipole interactions)

(*c*)　forces arising from the interaction between molecules having dipoles and those in which dipoles are induced by neighbouring molecules (dipole-induced dipole interaction)

(*d*)　forces arising from the interaction between ions and molecules having dipoles induced by the ions (ion-induced dipole interactions)

(*e*)　forces between neutral atoms or molecules (dispersion forces)

(*f*)　hydrogen-bonding interactions

Interactions of types (*b*) (*c*) and (*e*) are often known collectively as Van der Waals' forces.

It is not necessary to have a detailed quantitative knowledge of these forces, but some appreciation will help in our understanding of chromatographic processes.

(*a*)　Coulombic forces between ions of opposite charge give rise to strong interactions and are found in ion-exchange chromatography where their presence is characterised by slow exchange of ions and hence relatively broad chromatographic peaks which are associated with a low column efficiency (see Section 3.4.2).

(*b*) Dipole-dipole interactions. If a molecule contains atoms with different electronegativities (different electron attracting power) the electrons constituting the bonds between the atoms will be shared unequally, ie there is a separation of charge, eg chlorine is more electronegative than carbon so that the bond between carbon and chlorine is polarised

$$\overset{\delta+}{C} - \overset{\delta-}{Cl}$$

and the bond is said to exhibit polarity. Although the C—Cl bond itself is polar, a molecule as a whole may not be polar if it is symmetrical. Consider tetrachloromethane (carbon tetrachloride). We can represent the molecule and indicate the charge separation as shown. Since the molecule has tetrahedral symmetry, and only C and Cl atoms are involved, the effects of charge separation will cancel and the molecule as a whole is therefore non-polar (apolar).

The ability of an atom within a molecule to attract electrons to itself can be measured on the Pauling electronegativity scale. On this scale carbon has an electronegativity of 2.5 whereas chlorine has a value of 2.9. The higher the value, the more electronegative the atom.

SAQ 2.3a   Which of the following molecules would you expect to be *polar*? Pauling electronegativity values are C (2.5), H (2.1) and Cl (2.9):

(*i*)   benzene C_6H_6

(*ii*)   chlorobenzene C_6H_5Cl

(*iii*) *n*-butane $CH_3-CH_2-CH_2-CH_3$

(*iv*)   cyclohexane C_6H_{12}

Molecular dipoles cause molecules to attract each other, eg

(like charges repel but unlike charges attract) so that they become orientated in definite directions (just like compass needles in a magnetic field). However, the molecules are in continuous motion, which tends to break up any preferred alignment. This thermal motion increases as the temperature is raised and the orientation of the molecules is disturbed. Thus, dipole-dipole interactions *decrease* as the temperature *increases*.

(*c*) and (*d*) Dipole-induced dipole and ion-induced dipole interactions both arise from a similar cause. A charge on one atom or ion causes a shift of electrons in a second molecule producing a partial charge separation and hence a dipole movement. This is called an *induced dipole* and persists only in the presence of the inducing atom or ion. The size of the induced dipole depends on the *polarisability* of the molecule. Only those molecules with conjugated electron systems can experience an induced dipole.

SAQ 2.3b	Which of the following molecules could experience an induced dipole? (*i*) *n*-butane (*ii*) buta-1,3-diene (*iii*) cyclohexane (*iv*) benzene

SAQ 2.3b

(*e*) Dispersion forces. These are forces that are found in all molecules (including the noble gases). They arise from the instantaneous dipole in an atom due to the movement of the electron 'cloud' relative to the nucleus. A series of snapshots of, say, the argon atom would show the electron cloud firstly on one side and then on the other side of the nucleus. Over a macroscopic time period, these instantaneous dipoles would average to zero. These dipoles do not interact with those of other molecules to produce an attraction since there will be repulsion just as much as attraction and there is no time for the instantaneous dipoles to align with one other. However, there is an instantaneous dipole induced dipole interaction which produces an attraction.

(*f*) Hydrogen-bonding interactions. These are a special form of a dipole/dipole interaction. The hydrogen bond is the force of attraction between a partial positive charge on the hydrogen atom which is attached covalently to an electronegative element such as oxygen, nitrogen or fluorine, and a partial negative charge on another oxygen, nitrogen or fluorine.

$$\underset{X}{\overset{\delta-}{X}} \text{---} \underset{H}{\overset{\delta+}{H}} \cdots \underset{X}{\overset{\delta-}{X}} \text{---} \underset{H}{\overset{\delta+}{H}} \qquad \text{Where } X = O, N \text{ or } F$$

Thus, the hydrogen bond is a bridging bond. It is of variable strength, determined by the particular electronegative elements involved.

Hydrogen bonding can occur *intramolecularly*, ie involving one molecule only and giving rise to ring formation, or *intermolecularly*, ie between two or more molecules giving rise to association.

eg, hydroxybenzoic acid (an aromatic acid) undergoes intramolecular hydrogen bonding because of the close proximity of the oxygen atom of the $-C=0$ group and the hydrogen atom of the $-OH$ group thus

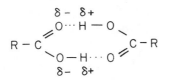

whereas an aliphatic acid can undergo intermolecular hydrogen-bonding thus:

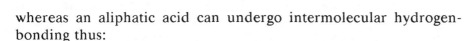

This has the effect of increasing the boiling point of the carboxylic acid in comparison with, say, an alcohol with the same relative molecular mass, M_r.

eg $\qquad CH_3CH_2OH$ ($M_r = 46$), bp $= 78\,°C$

$\qquad H.CO_2H$ ($M_r = 46$), bp $= 101\,°C$

It is for this reason that long chain fatty acid molecules are methylated to give the methyl ester

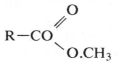

before chromatography by gc. Methylation removes the H atom which was responsible for the hydrogen bonding. Hydrogen bonding also has an important role to play in the solubility of one substance in another (or the solute in a solvent). Methanol/water mixtures are commonly used as the mobile phase in reverse phase liquid chromatography. Water itself is strongly hydrogen bonding, all the hydrogen atoms having a partial positive charge and the oxygen atoms bearing a partial negative charge.

The association of the water molecules in this way accounts for the high boiling point of water compared to its relative molecular mass.

If water and methanol are to mix, the methanol, which itself is associated, must be able to break some of the water hydrogen bonds and replace the water molecules by itself.

Hydrocarbons (eg alkanes) are non associated substances. In order to dissolve in water, hydrogen bonds between the water molecules would have, to be broken and reduced in number. With hydrocarbons this cannot occur since the hydrocarbons themselves do not contain —OH groups for hydrogen bonding.

∏   Would you expect a long chain alcohol, eg $C_{10}H_{21}OH$ to be soluble in water?

Molecules or groups which have an affinity for water are called *hydrophilic* (or *lipophobic*) whilst those with no affinity for water are called *hydrophobic* (or *lipophilic*). We have seen that the —OH group is hydrophilic but an alkane group is hydrophobic. Solubility is determined by the hydrophilic/hydrophobic (hydrophilic/lipophilic) balance within the molecule. Straight chain hydrocarbons with five or more carbons are too hydrophobic (lipophilic) and they are virtually insoluble in water.

To generalise, we can say that water tends to dissolve water-like molecules; nonpolar solvents tend to dissolve hydrocarbon like molecules. Polar solvents dissolve polar solutes but nonpolar solvents dissolve nonpolar (or moderately polar) solutes. Thus 'like dissolves like'.

SAQ 2.3c

Explain the following:

(*i*) *n*-pentane is insoluble in water but *n*-butyl alcohol and diethyl ether have about the same solubility (8 g in 100 cm³ water).

(*ii*) The boiling points of diethyl ether, *n*-pentane and *n*-butyl alcohol are 35 °C, 36 °C and 117 °C respectively.

2.3.2. Interactions Between Molecules

In gas chromatography there are only two sets of interactions to consider; those between the solute molecules and those between solute molecules and the stationary phase. In liquid chromatography the mobile phase (considered as being inert in gas chromatography) plays an important part in the separation process. We can represent the sets of interactions to be considered as:

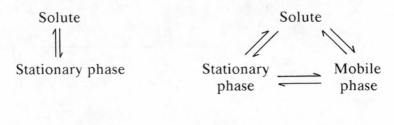

Gas chromatography   Liquid chromatography

To simplify our thinking, we need a guiding principle which we can apply when discussing the interactions between molecules. This principle is 'like has an affinity for like' and the yardstick we use is the polarity of a molecule, so that we can say polar molecules have an affinity for polar molecules and apolar molecules have an affinity for apolar molecules.

∏   Do you think polarity is a good measure of the potential for interaction of a molecule?

No, it is not, but it has the merit of being simple. It does not tell us anything about the dispersion forces, nor about the polarisability of a molecule or its ability to undergo hydrogen-bonding. Other measures of interaction (eg solubility parameters) can be left to a more advanced study.

(1) *Gas Chromatography*

We will consider the two extreme cases of (*a*) an apolar stationary phase and (*b*) a polar stationary phase.

(a) Apolar stationary phase

In this instance, the only interactions involved will be the London dispersion forces; coulombic, dipole and induced dipole interactions being absent. Differences in the dispersion forces will be reflected in the boiling points of the components of a mixture. With an apolar stationary phase, we can predict that apolar component molecules will be more strongly retained than polar component molecules, and of these two types of molecule (polar and apolar) those with the lower boiling points (higher vapour pressures) will be eluted before those of high boiling points.

(b) Polar stationary phase

With a polar stationary phase, the greatest affinity will be shown by polar solute molecules because these undergo dipole-dipole interactions in addition to the dispersion forces. Therefore, polar compounds are eluted more slowly and apolar compounds rapidly as they show little affinity for the stationary phase. Solute molecules which are polarisable can also show dipole-induced dipole interactions, and the retention of these molecules will depend on the degree of this interaction. With polar stationary phases, boiling-point considerations are often less important than the polar interactions in determining retention.

SAQ 2.3d

The following two chromatograms were obtained on (i) squalane (a hydrocarbon) and (ii) polyethyleneglycol (PEG) (a substance with an ether-like structure

$$-CH_2-CH_2-O-CH_2-CH_2-O-CH_2-CH_2-OH).$$
$$\longrightarrow$$

SAQ 2.3d
(cont.)

The three peaks are:

		Boiling Point	Polarity
A.	methanol	65 °C	high
B.	methyl ethanoate	57 °C	medium
C.	diethyl ether	36 °C	low

Assign the components A, B and C to the three peaks in the two chromatograms, and give your reasoning.

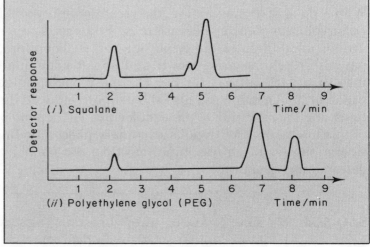

SAQ 2.3d

∏   Does ethanol have any additional interaction with PEG which would increase further its retention time?

Yes, hydrogen bonding

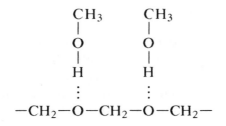

(2) *Liquid Chromatography*

In liquid chromatography, we have three sets of interactions to consider:

solute $\rightleftharpoons$ mobile phase,

solute $\rightleftharpoons$ stationary phase

mobile phase $\rightleftharpoons$ stationary phase.

To achieve a satisfactory separation, all three sets of interactions should be optimised.

∏   If one set of interactions predominates what would be the effect on the chromatography?

For a sample to be transported through the column it must be soluble in the mobile phase. If, however, interaction with the mobile phase is too strong, there will be little or no retention and the sample is eluted very rapidly. If the interaction with the stationary phase is too strong, retention times will be very long.

Stationary phase/mobile phase interactions are not usually very strong but we saw in Section 1.5.2 that the solubility of the stationary phase in the mobile phase caused problems with column stability which can be overcome by using chemically bonded phases (Section 1.5.3). There are instances where interaction between the mobile and stationary phase is required, eg when the mobile phase contains a molecule or ion that is to be retained on the stationary phase to form a secondary stationary phase of appropriate selectivity for the separation (eg in ion-pair chromatography) and ligand-complexation chromatography. However, this introduction will not deal with these techniques.

In order to obtain the correct balance of interactions in practise, it is more convenient to vary the mobile phase composition than to change the stationary phase, ie to use isocratic or gradient elution (Section 1.2.2). In gradient elution, the mobile phase composition is changed so that an *increase* in *solvent strength* causes a *decrease* in retention.

∏   Can you suggest what property we could use as a measure of solvent strength.

     Just as in gas chromatography where we used polarity in discussing solute molecules and the stationary phase, we can use polarity here to describe solvent strength.

In Section 1.5.3 we defined the terms *normal phase* and *reverse phase* chromatography, and it is important to recognise the difference between the two since our approach to changing solvent strength depends on the system we are using.

We still use our guiding principle 'like dissolves like' or 'like has an affinity for like'.

∏   The table below is incomplete. Try to complete it.

Type of chromatography	Polarity of stationary phase	Polarity range of mobile phase	Order of elution Polar/ non-polar	Effect of increase of polarity of mobile phase
Normal Phase				
Reverse Phase				

Our definition of *normal phase* was that the stationary phase is more polar than the mobile phase and our definition of *reverse phase* was that the stationary phase was less polar than the mobile phase so that the first two columns should read:

Type of chromatography	Polarity of stationary phase	Polarity range of mobile phase
Normal Phase	Polar	Weak to medium polarity
Reverse Phase	Non-polar	Strong to medium polarity

To determine the order of elution of two components, A and B, where A is more polar than B on the principle of like having an affinity for like in the normal phase system (polar stationary phase), the more polar molecule A will have the stronger affinity for the stationary phase and will be eluted after B. In the reverse phase system the less polar molecule B will have the greater affinity for the stationary phase and will be eluted after A.

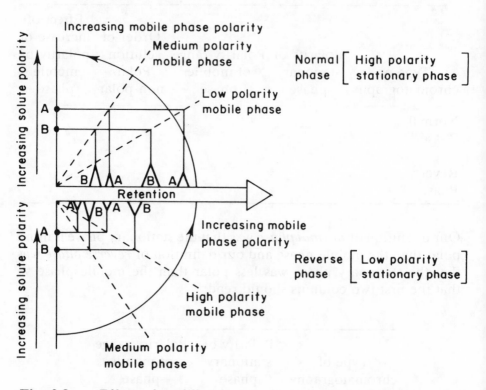

Fig. 2.3c.   *Effect of mobile phase polarity on retention in normal phase and reverse phased liquid chromatography*

To see the effect of changing the mobile phase polarity, refer to Fig. 2.3c. For the upper diagram (normal phase system), as we *increase* the mobile phase polarity we see that the retention times *decrease*. By increasing the polarity of the mobile phase we are making it more like the stationary phase, in terms of polarity, so that the mobile phase is better able to compete with the stationary phase for the solute molecules. The solute molecules will therefore spend more time in the mobile phase and be eluted more rapidly.

For the lower diagram (reverse phase system), an increase in mobile phase polarity will make it less like the stationary phase in polarity and therefore less able to complete with it for the sample molecules. The solute molecules will therefore spend more time in the stationary phase and the retention times will increase.

We can complete our table to give:

Type of chromatography	Polarity of stationary phase	Polarity range of mobile phase	Order of elution Polar/ non-polar	Effect of increase of polarity of mobile phase
Normal Phase	Polar	Weak to medium polarity	More Polar eluted last	Retention times decrease
Reverse Phase	Non-polar	Strong to medium polarity	More Polar eluted first	Retention times increase

We can also see from Fig. 2.3c that an increase in mobile phase polarity reduces the separation of the two components A and B in normal phase chromatography but increases the separation in reverse phase chromatography.

2.3.3. The Effect of Temperature and Flow Rate on Retention

The dependence of the distribution coefficient (K) with temperature (T) is given by the van't Hoff equation

$$\frac{d \ln K}{dT} = \frac{\Delta H}{RT^2} \qquad (2.25)$$

where ΔH is the enthalpy of solution of the solute molecules from the mobile phase to the stationary phase. If the phase ratio V_s/V_m

is independent of temperature, the capacity factor k' may be written for K, hence

$$\frac{d \ln k'}{dT} = \frac{\Delta H}{RT^2} \qquad (2.26)$$

This equation shows that retention time is inversely proportional to the square of the temperature, and, in gas chromatography particularly, the temperature of the column is an important way of controlling retention times. The enthalpy of solution of a solute transferred from the gas phase to the liquid phase is quite large, and a change of temperature of, say 30 °C, will have a pronounced effect on the retention time. In liquid chromatography, the solute is being transferred only from one liquid phase into another liquid (or pseudo liquid) phase, and the enthalpy change and hence the effect of temperatures will be much less. Liquid chromatography is therefore often carried out at ambient temperatures although slightly elevated temperatures are beneficial in improving column efficiency. There are also improvements in the reproducibility of k' values if the column is thermostatted.

In general, ΔH values are negative and the numerical values increase with increasing retention volume. For two components (1) and (2), it can be seen from the expression

$$\frac{d \ln(k_1'/k_2')}{dT} = \frac{\Delta H_1 - \Delta H_2}{RT^2} \qquad (2.27)$$

that the separation should detereorate if the column temperature is raised. However, this strictly only applies to ideal systems, and in practice there are many examples where an increase in temperature leads to an improved separation.

Eq. 2.28 relates the retention time of a compound to the mobile phase flow rate:

$$t_R = \frac{L}{\bar{u}_m}\left(1 + K\frac{V_s}{V_m}\right) \qquad (2.28)$$

so that an increase in the linear flow rate, $\bar{u}_m$, will decrease the retention time. As we shall see in Section 3.5 the flow rate effects the efficiency of the column.

In gas chromatography, because of the compressibility of the gas, changes in flow do not act instantaneously; the new flow rate takes time to establish itself which makes reproducible changes difficult to make. This, together with the cost of flow programming equipment for gases, means that changes of flow rate are rarely used in gas chromatography to control retention times.

In liquid chromatography, the affect of a change in flow rate on the efficiency of a column is less pronounced. There is no time lag when the flow rate is changed because the liquid is essentially incompressible and the equipment required to carry out the process is relatively cheap and readily available. Changes of flow rate (flow programming) may therefore be used in liquid chromatography to control retention times.

Learning Objectives

You should now be able to:

- understand the fundamental equations used in chromatography;

- define the retention parameters used in column, thin-layer and paper chromatography;

- be able to predict the retention behaviour of a substance in terms of the forces between the solute, stationary phase, and in liquid chromatography, the mobile phase.

3. The Quality of Chromatographic Separations

Overview

This section looks at the features of a chromatogram in terms of the peak shapes and widths, and the separation of the components of a mixture. The processes which lead to band broadening as the sample passes through the column and their effect on column efficiency are considered. The selectivity of the chromatographic process and the column efficiency will be brought together in a discussion of the resolution of chromatographic bands on the column and the methods used to optimise a separation are described.

3.1. INTRODUCTION

Take a look at a typical chromatogram:

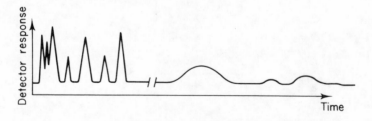

What do you notice about this chromatogram?

Firstly, the peaks are symmetrical and we can therefore describe them as having a Gaussian distribution or profile.

Secondly, the peaks which are eluted earliest are sharp but as time increases the later peaks become broader. Since the chromatogram is a plot of detector response *versus* time it means that for the earlier sharper peaks the concentration of a component as it goes through the detector is high and detection is easy. With the broader diffuse peaks the concentration of sample in the detector at any instant is low and detection is more difficult. The two cases are illustrated in Fig. 3.1a where the 'window' indicated by the vertical parallel lines represents the detector cell.

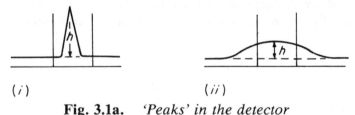

(*i*)   (*ii*)

Fig. 3.1a.   *'Peaks' in the detector*

In Fig. 3.1a(*i*) the peak is sharp and the detector response (proportional to h) is large, but in Fig. 3.1a(*ii*), the peak is diffuse, h is small and the detector response is small. However, the concentration of solute is proportional to the peak area, so that although Fig. 3.1a(*i*) gives the larger detector response if the band is narrow, the concentration may be quite small. On the other hand in Fig. 3.1a(*ii*) the band is broad and although h is small the total area may be quite large. Although peak height is sometimes taken as a measure of sample concentration, this is valid only for peaks of equal width; peak area is therefore a more reliable measure for quantitative analysis.

Thirdly, the peaks do not all take the same time to pass through the column, some being eluted very rapidly and others being retained on the column a considerable time.

Finally, the peaks at the beginning of the chromatogram are crowded together and are not well separated.

These four features are extremely important in chromatography and we shall look at them in some detail in the rest of Part 3.

3.2. PEAK SHAPE

Diffusion is a very commonly observed phenomenon. Any parcel of unbound molecules will tend to expand the volume that they occupy. In doing so the molecules become more diluted and increase their entropy or the randomness of their distribution. Thus, they are merely obeying the second law of thermodynamics. Although it is not necessary to the concept of diffusion, we usually think of the molecules moving from a region of high concentration to a region of low concentration. This presupposes that the molecules are 'aware' of high and low concentration regions. In fact the molecules will diffuse in all directions since their movement is completely random.

Diffusion

However, a region of low concentration of molecules will experience a net increase in flux and a region of high concentration a net decrease in flux.

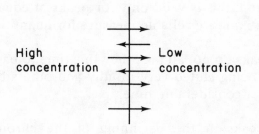

More molecules will flow from the high to the low concentration region than from the low to the high, and there will be a general levelling out of concentration.

The initial profile of the solute molecules can be represented by a sharp spike but the concentration profile as it evolves with time will be Gaussian in shape. (Fig. 3.2a.)

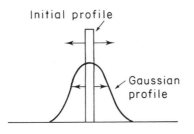

Fig. 3.2a. *Formation of a Gaussian profile*

3.3. SORPTION ISOTHERMS

The amount of a particular substance *sorbed* (or taken up) by the stationary phase depends on the concentration in the mobile phase. The relation between the amount sorbed and the concentration, in the mobile phase, at constant temperature, is called the *sorption isotherm*. The term *sorption* is a general description that covers several mechanisms including the process of *adsorption* (ie interaction of solute molecules with a solid surface), *partition* (dissolution of solute molecules in a liquid), ion-exchange (exchange of solute ions with ions held in the stationary phase and *size exclusion* (restriction of diffusion of solutes molecules through a porous stationary phase).

3.3.1. Sorption Isotherms and Peak Shape

Fig. 3.3a shows the form of the isotherm when the relation between the solute concentration in the stationary phase (C_s) and the solute concentrations in the mobile phase (C_m) is linear. The slope of the line is given by the ratio C_s/C_m.

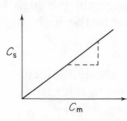

Fig. 3.3a.   *The linear sorption isotherm*

Π   What is this ratio called?

You should recognise it as defining the distribution coefficient K.

A chromatographic system where this relationship is linear is described as *linear chromatography* and will give rise to a Gaussian distribution of the solute along the column.

Fig. 3.3b shows a linear isotherm and Fig. 3.3c the characteristics of the chromatographic band with which it is associated.

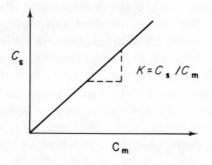

Fig. 3.3b.   *Sorption isotherm*

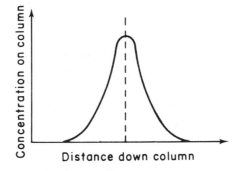

Fig. 3.3c. *Distribution of component on a chromatographic column*

In Section 1.4 we saw that the rate of migration of the solute band through the column is inversely proportional to the distribution coefficient. With a linear isotherm, K is independent of concentration (ie the slope is constant). Therefore, although Fig. 3.3c shows us that there is a concentration gradient across the peak, the maximum concentration being at the centre of the peak, the value of K will be constant across the peak. This means that the whole peak will move at the same rate and the basic shape will remain the same, ie Gaussian. As we have seen the peak will in fact broaden as it moves down the column due to the effects of diffusion, but for the present discussion we will ignore this. We are therefore talking about *linear ideal chromatography*, the *ideal* referring to the fact that diffusion has been ignored.

In adsorption chromatography, it is common to find a sorption isotherm that is concave to the C_m axis (often referred to as a Langmuir type isotherm) Fig. 3.3d. This behaviour arises when there are strong interactions between the solute and the stationary phase, but the solute/solute interactions are relatively weak. Initially, the amount of solute adsorbed onto the stationary phase increases rapidly as the concentration in the mobile phase is increased until a monolayer of solute molecules is formed on the adsorbent. Because the interactions between solute molecules are weak, the adsorption ceases with the formation of the monolayer and the amount adsorbed remains constant even though the concentration in the mobile phase in increased further. Hence the distribution coefficient K is large at low C_m values but decreases as C_m increases.

∏   Fig. 3.3e shows the peak shape associated with such an isotherm. Can you give an explanation of this peak shape?

[Hint – Remember that the rate of movement is inversely proportional to K]

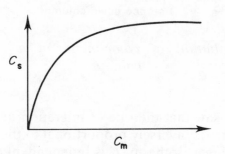

Fig. 3.3d.   *A Langmuir type isotherm*

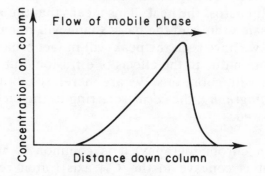

Fig. 3.3e.   *The peak shape associated with a Langmuir type isotherm*

K is given by the slope of the tangent to the isotherm at any value of C_m thus:

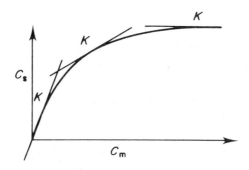

This means that the rate of migration is not now the same across the band. If we represent the rate of migration by arrows, the edges of the band, where the concentration is lowest, have the highest K values and therefore move relatively slowly compared to the centre of the band which has a higher concentration and hence lower K value. The centre of the band therefore moves more rapidly catching up with the front and leaving the tail behind.

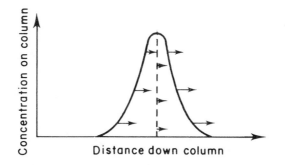

The resulting solute peak has a sharp front and a long diffuse tail (Fig. 3.3e) and is described as a *tailing peak*.

In partition systems, the isotherm is usually convex to the C_m axis (an anti-Langmuir type isotherm). This isotherm type arises when the interactions between solute molecules are strong but those between the solute and the stationary phase are relatively weak. Thus, at low concentrations, the solubility of the solute molecules is low and K has a low value, but once a few solute molecules are dissolved in the stationary phase, the strong solute/solute interactions draw further molecules into the stationary phase and K increases rapidly.

∏   Following the previous explanation, what peak shape would you expect from a partition system?

A peak with a diffuse front and a sharp tail is described as a *fronting peak*.

You should have reasoned that with an anti-Langmuir type isotherm K will increase as the concentration increases across the band, so that it will be lowest at the edges and highest at the centre of the peak. The rate of migration will therefore be lowest at the centre and highest at the edges. The tail will tend to catch up with the centre and the front will move away from the centre giving a peak with a sharp tail and a diffuse front ie a *fronting peak*. This process is represented in Fig. 3.3f-h.

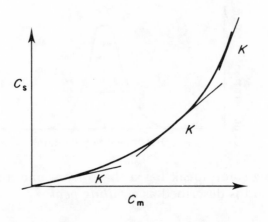

Fig. 3.3f.   *Anti-Langmuir isotherm (K increases with C_m)*

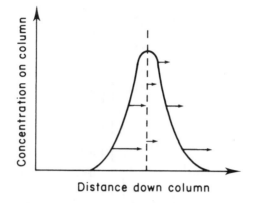

Fig. 3.3g. *Rate of movement greatest at edges of peak*

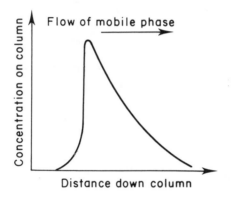

Fig. 3.3h. *A fronting peak*

Although you may have deduced the peak shape given in Fig. 3.3h the equation for the elution curve at finite concentration in fact gives a curve in which the leading edge is convex and not concave as in Fig. 3.3h. Fig. 3.3i shows a series of fronting peaks for differing sample sizes and Fig. 3.3j shows the isotherm that can be calculated from the diffuse edge of the elution peak.

Several other factors may also give rise to an asymmetric peak, in particular tailing due to slow injection onto the column but these can be limited by proper attention to the operating conditions.

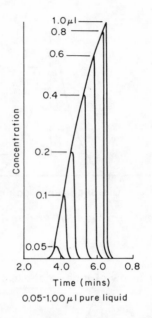

Fig. 3.3i. *Chromatograms of n-octane on 20% H_2O (w/w)-Porasil D as a function of liquid sample sizes; T = 12.6 °C*

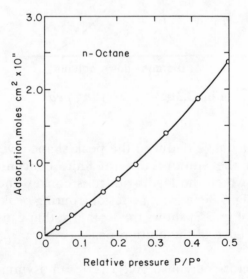

Fig. 3.3j. *Adsorption isotherms for 20% H_2O (w/w)-Porasil D; T = 12.5 °C*

In general, tailing and fronting peaks are broader than the corresponding Gaussian shaped peak. In Section 3.7.1 we shall see that broad peaks lead to poorer resolution for two compounds and are to be avoided.

∏   Can you suggest how this can be done?

Both forms of the isotherm can be regarded as being linear if the solute concentration is small enough. Thus, if the amount of sample injected onto a column is kept low, Gaussian peaks will usually be obtained from a non-linear isotherm. In gas liquid chromatography for a typical packed column (4mm id) the sample size of a given component in a mixture should be kept below 0.1 mg (0.1 μl if the M_r = 100). In gas solid chromatography, the sample size must be considerably less if Gaussian peaks are to be obtained. When the sample size is too large and the peaks become distorted, the column is said to be *overloaded*.

3.3.2. Peak Asymmetry Factor

The simplest way of measuring the degree of peak distortion (skew) is by the peak asymmetry factor (A_s) which is normally measured as shown in Fig. 3.3k.

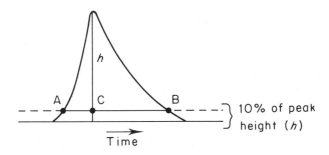

Fig. 3.3k.   *Measurement of Peak Asymmetry*

$$A_s = \frac{CB}{AC} \tag{3.1}$$

The lengths of AC and CB are measured at a point corresponding to 10% of the total peak height. Values of the asymmetry factor greater than 1.2 indicate that the quality of the column packing is poor and a column giving a value greater than 1.6 should be discarded since it will be difficult to get the required resolution of the components of a mixture.

3.3.3. The Effect of Peak Asymmetry on Retention Times

The effect of peak asymmetry on the retention time is illustrated in Figs. 3.3l (*i*) and (*ii*).

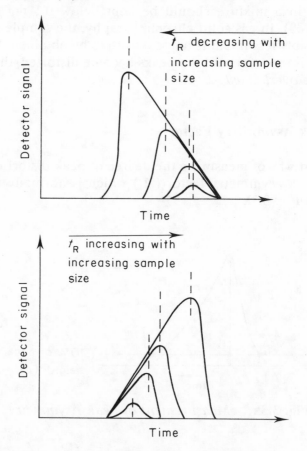

Fig. 3.3l. *(i) and (ii) Effect of peak asymmetry on retention time*

In the case of the Langmuir isotherm which gives rise to a tailing peak (Fig. 3.3l(i)), the speed of movement of the zone increases as the concentration increases so that the larger the sample size the smaller the retention time measured to the peak maximum. For an anti-Langmuir type isotherm, which gives rise to the fronting peak (Fig. 3.3l(ii)), the speed of movement of the zone decreases as the concentration increases, and the retention time increases with sample size. The small symmetrical peak at the end of the tail in a tailing peak and at the front of the fronting peak arises from a linear (or pseudo-linear) portion of the isotherm; the value of K for these peaks will be constant and hence the retention time will be constant.

Thus, not only do asymmetric peaks give poorer resolution of the components in a mixture but they also introduce an uncertainty into the measurement of retention times and retention volumes.

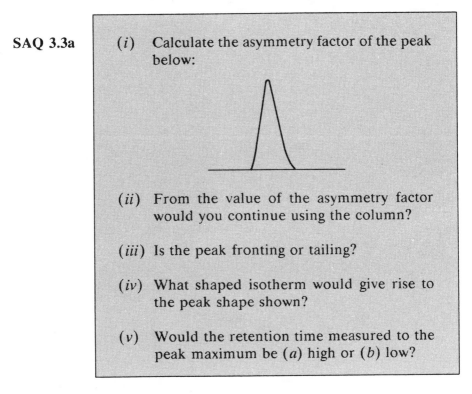

SAQ 3.3a

(i) Calculate the asymmetry factor of the peak below:

(ii) From the value of the asymmetry factor would you continue using the column?

(iii) Is the peak fronting or tailing?

(iv) What shaped isotherm would give rise to the peak shape shown?

(v) Would the retention time measured to the peak maximum be (a) high or (b) low?

SAQ 3.3a

3.4. COLUMN EFFICIENCY

The second point that we noticed about our chromatogram was that the peaks at the start are sharp, whilst the later ones become broader and more diffuse. The sharpness of the peak is a property of the chromatographic column and is described as the *column efficiency*. In Section 1.4 we defined a theoretical plate as being a section of the column in which the sample equilibrated between the stationary phase and the mobile phase. The flow of a fluid through a packed bed is a very complicated process, and the mechanisms of band broadening affecting a solute as it passes through the column do not lend themselves to an exact mathematical treatment. However, predictions made from theory are in good agreement with the practical results so justifying the attempts made to explain the complex phenomena.

3.4.1. Measurement of Column Efficiency

The extent of band broadening determines the efficiency of the column and each process responsible for this broadening contributes to the overall width of the band. The contribution from each process can be described mathematically in terms of the *variance* (σ^2) of the Gaussian chromatographic peak which is the square of its *standard deviation* (σ). In chromatographic terms, the retention volume would correspond to the mean so that the variance is a measure of deviations from this mean. Thus for a *single* component not all particles elute at a time t_R to give a retention volume V_R. Some particles elute in a fractionally shorter time and some in a longer time. This is shown by the Gaussian shaped elution peak that we get in linear chromatography where the basal peak width (W_b), ie the width of the peak at the base line, is related to the standard deviation by $W_b = 4\sigma$, (Fig. 3.4a).

The efficiency of the column can be expressed as the number of theoretical plates or plate number (N) or by the *height equivalent to a theoretical plate* (HETP or simply H).

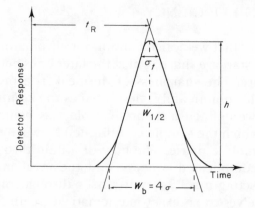

Fig. 3.4a.   *Measurement of N*

The column efficiency can be measured from the peak profile (Fig. 3.4a) in several different ways, but the most commonly used expressions are:

(1) $$N = \left(\frac{t_R}{\sigma_t}\right)^2 \qquad (3.2)$$

(2) $$N = 5.54\left(\frac{t_R}{W_{\frac{1}{2}}}\right)^2 \qquad (3.3)$$

(3) $$N = 16\left(\frac{t_R}{W_b}\right)^2 \qquad (3.4)$$

where

σ_t^2 = peak variance in time units (peak width at $0.882h$)

$W_{\frac{1}{2}}$ = peak width in time units at half peak height

W_b = peak width in time units at the base line

t_R = uncorrected retention time

The basal peak width is obtained by drawing tangents to the points of inflexion of the Gaussian profile as shown. Since N is dimensionless, t_R and the peak width must have the same units of measurement (eg *both* in time units or *both* in distance on the chromatogram).

∏   Which measure of N do you think would be the best to use, and why?

All three methods require the measurement of t_R, so that any differences must be in the measurement of peak width. All three measurements need to know the position of the base line; the first two to give 'h' and the third because the width of the peak is measured at the base line. Measurement of σ_t can be made either as the peak width at $0.882h$ or from the basal peak width. If σ_t is measured at $0.882h$ it is subject to large error since the peak width is very small at this point. Measurements depending on W_b rely on the accurate positioning of the tangents. The measurement of $W_{1/2}$ still requires a knowledge of h but once determined the points of measurement are defined by the elution curve. These factors point to the expression

$$N = 5.54 \left(\frac{t_R}{W_{\frac{1}{2}}} \right)^2$$

as being the best method for determining N.

Since N is directly proportional to column length an increase in length will always lead to a better separation because of the increased plate number. The proportionality of N and L is given by the equation

$$N = L/H \qquad\qquad (3.5)$$

Small values of H, therefore, mean large values of N and high efficiency. Efficiencies can also be quoted in terms of H rather than N because H is independent of column length.

SAQ 3.4a   The following chromatogram was obtained from
a 2 metre packed column. Calculate the values
of N and H for the second peak using two different methods. Comment on any difference you
get.

Note that the first peak is an unretained component.

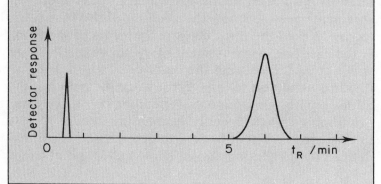

3.4.2. Effective Plate Number (N_{eff})

In Section 2.1.3 we saw that the adjusted retention volume took into account the time that solute species spend in the mobile phase, whereas the total (uncorrected) retention time includes this. In measuring the plate number, we have again used the uncorrected retention time so that the value of N overestimates the efficiency. The *effective plate number* (N_{eff}) takes this into account, ie

$$N_{eff} = 5.54 \left(\frac{t_R - t_m}{W_{\frac{1}{2}}} \right)^2 \qquad (3.6)$$

and $$H_{eff} = L/N_{eff}$$

Effective plate number and plate number are related by the equation

$$N_{eff} = \left(\frac{k'}{1 + k'} \right)^2 N \qquad (3.7)$$

As the retention time increases, the value of N_{eff} approaches that of N, ie it is for the earlier eluting peaks that N_{eff} is a better indicator of column efficiency.

SAQ 3.4b	From the chromatogram in SAQ 3.4a calculate the values of N_{eff} and H_{eff} for the second peak.

3.5.   *BAND BROADENING* PROCESSES

Although mathematically complex, a qualitative description of the processes which lead to band broadening is not difficult, but you could if you wished, take Section 3.5 as optional.

There are three main contributions to band broadening: *eddy diffusion* (or the *multiple path* effect), *longitudinal molecular diffusion* and *mass transfer*. Whilst we can recognise all three processes in both gas and liquid chromatography, the relative importance of each is not the same in the two techniques. This is due to the large difference in physical properties such as diffusion coefficients and viscosity between gases and liquids. In gas chromatography, eddy diffusion, longitudinal molecular diffusion and mass transfer effects in the stationary phase are important whilst in liquid chromatography, molecular diffusion is often ignored but mass transfer in both the stationary and mobile phases is of importance.

The variances for each of the band broadening processes are additive to give the overall variance for the system, and this is a measure of column efficiency. It is expressed in terms of H (the plate height) rather than σ^2, the two being related by the column length, L.

The overall value of H is a function of the average linear velocity of the mobile phase, $\bar{u}$, the general equation being expressed in the form:

$$H = \left(\frac{1}{A} + \frac{1}{C_m \bar{u}} \right)^{-1} + \frac{B}{\bar{u}} + C_s \bar{u} + C_{sm} \bar{u} \quad (3.8)$$

where each term represents a contribution to band broadening due to diffusion and/or mass transfer effects.

3.5.1. The *A* term: Multiple Path Effect (Eddy Diffusion)

The flow pattern through a bed of a granular material is very tortuous as solute and mobile phase species take paths of least resistance to fluid flow. The velocity of a single particle through the packed bed will fluctuate between wide limits, and the total distances travelled by individual particles will also vary. These fluctuations are random because the structure of the bed which causes them is random. Particles travelling through open pathways will move more rapidly than those in narrow pathways. The simple theory of eddy diffusion assumes that a particle will remain in a single flow path. In practice, this is not the case, since there is nothing to stop it from diffusing laterally from one flow path to another. This process (called 'coupling') averages out the two flow paths and reduces the amount of band broadening so that the final band width, although still greater than the initial band width, is less than if coupling had not occurred.

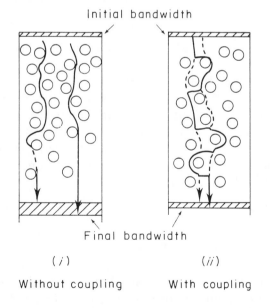

Fig. 3.5a. *Multiple path effect (eddy diffusion)*

Fig. 3.5a(i) shows the path of two particles locked into their separate flow paths. Fig. 3.5a(ii) shows how a particle may diffuse laterally (shown by the horizontal lines) from one flow path to another. The dotted lines indicate the flow path the particle would have taken without coupling.

The contribution to the overall-plate height, H, from the multiple path effect is given by:

$$A = 2\lambda\, d_p \tag{3.9}$$

where λ = a packing constant (≈ 0.5 for a well packed column) and d_p = the particle diameter.

3.5.2. The B Term: Longitudinal (Molecular) Diffusion

As a solute band moves through the column, diffusion in the direction of flow (longitudinal or axial) will also occur. This diffusion not only occurs in the fluid phase (ie gas or liquid) but to a much lesser extent at interfaces between surfaces (eg on the surface of a solid adsorbent). The plate height contribution of longitudinal diffusion is given by:

$$B = 2\,\gamma\, D_M \tag{3.10}$$

where γ is an obstruction factor which accounts for the fact that diffusion is hindered by the column packing. For packed columns, typical values are 0.6–0.8 and for capillary columns 1.0. D_M is the coefficient of diffusion of the solute species in the mobile phase. Because diffusion coefficients in liquids (typically $\approx 10^{-5}$ cm^2 s^{-1}) are much smaller than in gases (typically 10^{-1} cm^2 s^{-1}) the effect of longitudinal molecular diffusion is often ignored in liquid chromatography.

3.5.3. The C Terms: Mass Transfer

Mass transfer relates to the rates at which solute species are sorbed and desorbed, and diffuse within each phase. This rate is controlled by two mechanisms:

(*i*) Sorption-desorption kinetics

This relates to the intermittant capture and release of solute species by the stationary phase and is typified by adsorption/desorption from surfaces when a molecule can detach itself only if it possesses sufficient energy (the activation energy) to rupture the chemical or physical bonding.

(*ii*) Diffusion controlled kinetics

This may originate in either a liquid stationary phase or in the mobile phase. Diffusion in the stationary phase is relatively simple to treat quantitatively but diffusion in the mobile phase is very complex.

Irrespective of the mechanisms involved it is convenient to divide the mass transfer effects into stationary and mobile phase mass transfer terms, C_s and C_m.

C_s Term: Stationary Phase Mass Transfer

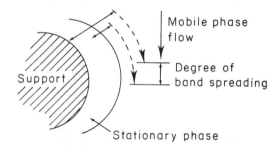

Fig. 3.5b. *Stationary phase mass transfer*

The rate at which solute species transfer into and out of the stationary phase makes a significant contribution to band broadening. The rate is controlled mainly by diffusion in a liquid stationary phase or by adsorption-desorption kinetics for a solid stationary phase. Because of statistical fluctuations, individual solute species will reside in/on the stationary phase for differing lengths of time. Those which spend most time in/on the stationary phase will, when they regain the mobile phase, have been left behind and the solute band will have been broadened. Mathematically a liquid and a solid stationary phase may be treated similarly in that in both cases a solute particle will spend a given time in the stationary phase. For a liquid this will depend on the rate of diffusion (D_s) and for a solid on the mean desorption time (t_d) (the mean time that a particle remains attached to the surface) which will in turn depend on the activation energy for the adsorption process.

The contribution to the overall plate height, H, for each case is given by:

(i) adsorption,
$$2\,t_\mathrm{d}\;\frac{k'}{(1\,+\,k')^2} \qquad (3.11)$$

(ii) partition,
$$\frac{d_f^2}{D_\mathrm{s}}\;\frac{k'}{(1\,+\,k')^2} \qquad (3.12)$$

where k' is the capacity factor and d_f the liquid film thickness. In adsorption it is important to ensure an energetically homogeneous surface to reduce H, whilst in partition, the liquid film thickness should be small and the liquid stationary phase should be chosen (where possible) to give high solute diffusion coefficients. Film thickness is controlled both by the amount of liquid stationary phase used (10% liquid/support material is common) and by the surface area of the support material.

C_m and C_{sm} Terms: Mobile Phase Mass Transfer

Mobile phase mass transfer may be divided into contributions from (a) the moving mobile phase and (b) the 'stagnant' mobile phase.

(a) *'Moving' mobile phase* (Fig. 3.5c).

This refers to the fact that solute species in the same flow path will not all move with the same velocity. One reason for this is that those in the centre will move more rapidly than those at the column walls which are influenced by frictional forces (the same effect can be observed if two pieces of wood are thrown into a stream; one in the centre of the stream will move faster than one close to the banks).

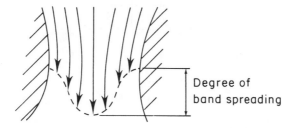

Degree of band spreading

Fig. 3.5c. *'Moving' mobile phase mass transfer*

Thus, the band is broadened. A rigorous treatment of all the processes occurring in a fluid flowing through a packed bed is not possible as they involve flow inequalities due to eddy diffusion and lateral mass transfer by diffusion and by convection. The contribution to the overall plate height, H, is given by:

$$C_m = \Omega d_p^2 / D_M \tag{3.13}$$

where Ω is a function of the packing structure, d_p the particle diameter and D_M is the coefficient of diffusion of the solute species in the mobile phase.

(*b*) *'Stagnant' Mobile Phase* (Fig. 3.5d).

If the stationary phase involves a porous material, the pores (ie the intraparticle void volumes) are filled with mobile phase at rest and there is very little interchange between this 'stagnant' mobile phase contained in the pores and the moving mobile phase in the outside of the pores (ie the interparticle void volume). To reach the stationary phase, solute particles will have to diffuse into this stagnant pool of liquid. Just as in the stationary phase itself some particles will diffuse deeper into the stagnant pool than others and will consequently be left behind, when they finally join the main stream again a broadening of the chromatographic band results.

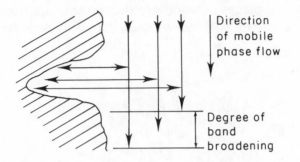

Fig. 3.5d.   *'Stagnant' mobile phase mass transfer*

The contribution to the overall plate height, H, from this stagnant mobile phase effect in porous spherical particles is

$$C_{sm} = \frac{(1 - \phi + k')^2 \, d_p^2}{30(1 - \phi)(1 + k')^2 \, \gamma \, D_M} \tag{3.14}$$

where ϕ is the fraction of the total mobile phase in the intraparticle space.

3.5.4.   The Overall Plate Height Equation

Eq. 3.8 has been used to enable chromatographers to understand and predict the effects of these various parameters in column performance in order to minimise the plate height H. Not all the physical quantities are of equal importance and some are more easily controlled than others. In general, we can say that a minimum value of H is obtained if we use thin liquid films (in gas chromatography), columns packed with small diameter particles, and a mobile phase of low viscosity. Although temperature does not appear in Eq. 3.8 it affects column efficiency through the effect it has on the rates of diffusion, on viscosity and on sorption/desorption kinetics, so that an increase in temperature leads to an increase in efficiency and a decrease in H.

In gas chromatography, the mobile phase mass transfer terms are of negligible proportions so that Eq. 3.8 reduces to:

$$H = A + \frac{B}{\bar{u}} + C_s \bar{u} \qquad (3.15)$$

This is the abbreviated form of the well known equation referred to as the van Deemter equation.

Fig. 3.5e shows the variation in plate height with the mean linear velocity of the mobile phase for typical values in gas and in liquid chromatography.

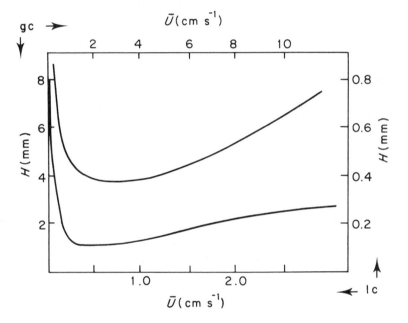

Fig. 3.5e. *Variation of plate height with linear velocity*

Both curves have the same form showing a minimum value of the plate height (H_{min} at a given flow rate ($\bar{u}_{min}$). Below this velocity, H is strongly dependent on diffusion effects (the B term) and at higher flow rates on the mass transfer terms (C).

The lc curve shows a flatter rise of H with velocity compared to the gc curve. This is due to the coupling affect which is more important in lc then in gc. The slow rise of H with $\bar{u}$ in liquid chromatography means that the flow rate is not so important in determining column efficiency, and higher flow rates can be used (and hence faster analysis achieved) without too much loss in efficiency. In practice the value of $\bar{u}_{min}$ tends to be rather low and leads to overly long analysis times so that a value $2 \times \bar{u}$ is often used.

In Section 1.2 we saw that capillary columns were narrow bore tubes with the stationary phase coated onto the inside walls. They contain no packing so that the multiple path term (A) is inappropriate in describing peak broadening, and the obstruction factor (γ) in the B-term is unity. However, mass transfer effects in the gas phase at right angles to the direction of flow must be considered. The plate height can be described by

$$H = \frac{B}{\bar{u}} + C_s \bar{u}$$

or specifically by the Golay equation

$$H = \frac{2DM}{u_O} + \frac{11k'^2 + 6k' + 1}{24(1 + k')^2} \frac{r^2 u_O}{D_m} + \frac{2k'(d_f^2 u_O)}{3D_s(1 + k')^2}$$

where r is the column radius, u_O the mobile phase velocity at the column outlet, D_s and D_m are the coefficients of diffusion for a solute in the stationary and mobile phases respectively and k' is the solute capacity factor. The second and third terms relate to mass transfer in the mobile and stationary phases respectively.

3.5.5.   Extra-column Band Broadening

Band broadening can occur in other parts of the chromatographic system as well as in the column, but in gas chromatography these effects are negligible. This *extra-column* band broadening can occur in the injector system, in connecting pipework and in the detector, and a well designed chromatograph should aim to keep the volumes in these three regions as small as possible. One of the most sensitive areas to band broadening is the connecting pipework between the column and the detector. The contribution to the plate height in an open tube is given by:

$$\sigma^2 = \pi\, r^4.F.L/24D_M \qquad (3.16)$$

where the symbols are as previously defined.

It is therefore necessary to use short, narrow bore tubes. The length should not exceed 30 cm and the internal diameter is usually 0.01 inch; narrow tubing tends to obstruct and wider tubing contributes significantly to band broadening. If the detector is not to broaden bands, the void volume of the cell should not exceed 10 μl for analytical applications, but where high efficiency columns are used the detector volume should be even smaller.

Sample introduction can also lead to band broadening. Although on-column injection through a septum gives the highest efficiency and is used extensively in gas chromatography, for liquid chromatography, where the injection is made against a higher pressure than in gas chromatography, problems occur with syringe injection. Injection with a valve employing a fixed volume sample loop is preferable even though the sample volume may contribute to band broadening because of slow removal of traces of sample from the walls of the valve.

3.6. SELECTIVITY AND RESOLUTION

The third feature of our chromatogram at the beginning of Part 3 is that not all the solutes have the same retention time, otherwise there would be no separation into individual component bands. We can say that compounds in the mixture were selectively retained or that the column shows *selectivity* toward the compounds of the mixture.

Selectivity is measured by the *separation factor* or *relative retention* (α) and may be given by any of the following relation:

$$\alpha_{21} = \frac{K_2}{K_1} = \frac{k'_2}{k'_1} = \frac{t'_{R_2}}{t'_{R_1}} = \frac{t_{R_2} - t_m}{t_{R_1} - t_m} \qquad (3.17)$$

The relation between the separation factor and the distribution co-efficients K_2 and K_1 of the two components (2) and (1) emphasise the thermodynamic basis for the separation. If (ΔG) is the difference in the standard free energies of the distribution of the solute between the stationary and mobile phase then

$$\Delta (\Delta G^{\ominus}) = -RT \ln \alpha \qquad (3.18)$$

The separation factor is usually measured from the ratio of the capacity factors (k'_2 and k'_1) or the adjusted retention times (t'_{R_2} and t'_{R_1}).

SAQ 3.6a

SAQ 2.3d shows the separation of methanol, methyl ethanoate and diethyl ether on squalane and polyethylene glycol. Given that $t_m = 0.3$ min:

(*i*) calculate the separation factors for the component pairs: methyl ethanoate/methanol; methyl ethanoate/diethyl ether and diethyl ether/methanol on the two columns.

(*ii*) on the basis of the α values, which column would you use to separate each pair of compounds?

In the SAQ we saw that the α values could be greater or less than unity. Using the definition $\alpha = K_2/K_1$, values of α would always be greater than unity. However in comparing the selectivity of columns it is common to give all the α values relative to a particular solute (or standard). When this is done it is almost certain that some values will be less than unity. This idea of using a 'standard solute' is used in another way. The retention time, being an absolute value, is subject to errors in measurement. If a standard solute is injected with a mixture each time and values of the retention relative to this standard (α_{ST}) are calculated, the errors tend to be self-cancelling and more reproducible retention data are obtained.

∏   Which of the following descriptions of α are correct?

It is

(a) a relative retention
(b) a distribution coefficient
(c) the separation factor
(d) the ratio of two distribution coefficients
(e) the ratio of two separation factors
(f) the ratio of two unadjusted retention times
(g) the ratio of two adjusted retention times
(h) the selectivity
(i) the capacity factor
(j) the ratio of two capacity factors
(k) the ratio of two peak widths
(l) a measure of the sorption free energy difference

We saw in the test that α can be described in several ways. Thus (a), (c), (d), (g), (h), (j) and (l) are all descriptions of α.

(a) and (c) are alternative names for α, which is a measure of the selectivity of the column, so that (h) also describes α. Remember that α is a *relative* value so that (d) and (g) are both definitions whereas (b) and (i) are not since they are not relative values. Finally, (l) relates α to the fundamental process of sorption in the chromatographic system.

3.6.1. Resolution

We have seen that the selectivity of the chromatographic system describes the separation of the band centres (as measured by the relative retention, ie by t_{R_1} and t_{R_2}), but it does not describe the separation of the two peaks adequately because this is also a function of the peak widths.

Thus, in Fig 3.6a we see that (i) represents the 'ideal' situation with the band centres well separated and the peak widths narrow. In (ii), although the band centres have the same separation, the peak widths are now very broad so that the peaks overlap and separation is incomplete. In (iii) the peak widths are the same as in (ii), but in order to achieve complete separation, the separation of the band centres must be increased. To do this the retention time of the second peak has been increased thus increasing the overall time for the analysis.

Band centres

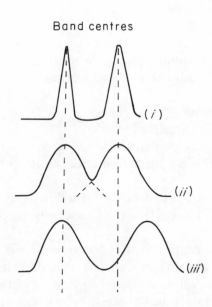

Fig. 3.6a. *Effect of peak width and relative retention on resolution*

The peak widths are a function of the column efficiency as measured by the number of theoretical plates (Section 3.4.2). Thus, the separation of two peaks is determined by the 'selectivity' of the stationary and mobile phase and also by the efficiency of the column. These two factors are combined to give the *resolution* (R_s) between two peaks, which is defined by:

$$R_s = \frac{t_{R2} - t_{R1}}{\frac{1}{2}(W_1 + W_2)})$$ (3.19)

where t_{r2} and t_{r1} are the retention times and W_1 and W_2 are the basal peak widths of the peaks 1 and 2 (see Fig. 3.6b). If the peak widths W_1 and W_2 are very similar Eq. 3.19 approximates to

$$R_s = \Delta t_r / W$$ (3.20)

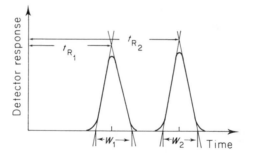

Fig. 3.6b. *Measurement of resolution*

SAQ 3.6b

Assuming that the two peaks can be represented by two triangles, calculate the value of R_s which corresponds to the situation where the two peaks are *just* separated at the base-line.

SAQ 3.6b

∏   In the answer to SAQ 3.6b we saw that baseline resolution was achieved when $R_s = 1.0$. In the two cases given below would we have a value of R_s greater or less than 1?

(a)

(b)

Clearly in example (a) $R_s > 1$ since $\Delta t_R > \frac{1}{2}(W_1 + W_2)$

and in (b) $R_s < 1$ since $\Delta t_R < \frac{1}{2}(W_1 + W_2)$

∏   If we now assume the peaks to be Gaussian in shape what effect will this have on the R_s value required to give base line separation; will the value of R_s need to (a) increase or (b) decrease?

[Hint. refer back to Fig. 3.4a]

We have calculated an R_s value of 1.0 for base line separation with triangular peaks, but Fig. 3.4a shows that a Gaussian peak is broader at the base line than the equivalent triangular shaped peak since it approaches the base line exponentially. (It is also more rounded at the peak maximum (see Fig. 3.6c).)

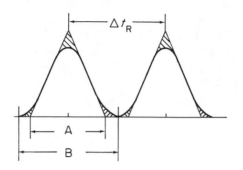

Fig. 3.6c. *Resolution of triangular and Gaussian shaped peaks*

This means that R_s must be greater than 1.0 for base line resolution since the peak width of the Gaussian peak (as measured by B) is greater than that of the triangular peak (as measured by A) but Δt_R is the same.

In practise, a value of $R_s = 1.0$ corresponds to a separation of the order of only 94%, ie only 94% of the peak is pure, the other 6% would be contaminated with the overlapping component. Base line resolution with Gaussian peaks is considered to be achieved with an R_s value of 1.5. However, as we shall see in the next section, it is not always necessary to achieve that degree of resolution in a separation.

Resolution is a dimensionless number and therefore the units of retention time and peak width must be the same. You will be quite familiar with measuring retention in units of time (ie min or sec) but not, perhaps, with measuring distance in time units. This should not give rise to any difficulty if you think of the basal peak width as being that time that a band of substance takes to pass through the detector. Since the chromatogram is recorded on a chart recorder running at a constant speed, the peak width measured as a distance on the chart paper and the peak width measured in units of time are related by the recorder chart speed (cm s^{-1}).If, as often happens, the chart is run at 1 cm s^{-1} then these two measurements are numerically the same.

SAQ 3.6c

The following two chromatograms represent the separation of a 2:1 mixture of cyclohexane and benzene on two different columns, column 1 being apolar and column 2 polar. For column 1, the chart speed was 1 cm s^{-1} and for column 2 it was 0.5 cm s^{-1}. All other chromatographic conditions were the same in each case.

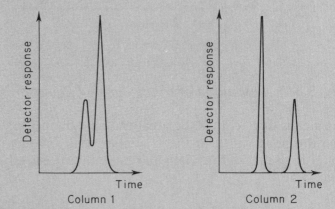

(*i*) Calculate the values of R_s for the two columns and comment on the results you get.

(*ii*) Assuming that the detector response is proportional to concentration, how do you account for the change in the elution order of cyclohexane and benzene on the two columns?

SAQ 3.6c

3.6.2. Estimation of Resolution

Although we can measure the resolution as described, this may not be necessary. One approach to *estimating* resolution is based on a standard set of resolution curves. By comparing your chromatogram with the standard curves, an approximate value of the R_s can be obtained without recourse to retention time and peak width measurements. Such a method gives a quick, convenient way to estimate resolution, but since the appearance of the chromatogram will depend on the relative peak heights of the two components whose resolution is to be measured, it is at best only an approximation. Fig. 3.6d (i) to (iv) show the standard resolution curves for peak height ratios of $1:1$, $4:1$, $8:1$ and $16:1$ with R_s values of 0.4 up to 1.25.

In (i) (peak height ratio $1:1$) it can be seen that an R_s value of 0.6 clearly gives two peaks, although the separation is poor, but in (iv), with a band ratio of $16:1$, the second peak appears only as an asymmetry on the tail of the first peak. These diagrams indicate clearly the effect of peak height ratios on the resolution required before the appearance of the minor peak becomes obvious.

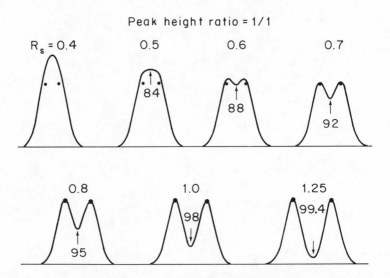

Fig. 3.6d. *Standard resolution curves for (i) a peak height ratio of 1/1 and R_s values of 0.4–1.25*

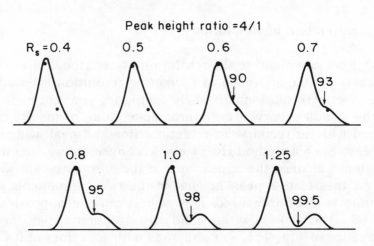

(ii) a peak height ratio of 4/1 and R_s values of 0.4-1.25

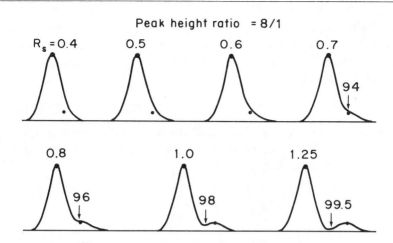

(iii) a peak height ratio of 8/1 and R_s values of 0.4-1.25

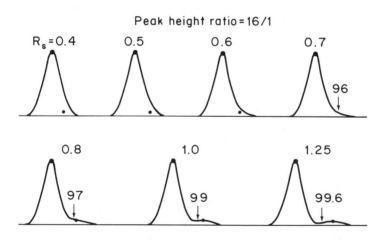

(iv) a peak height ratio of 16/1 and R_s values of 0.4-1.25

In the case of overlapping peaks, the positions of the true band centres of the two components will not correspond with the apparent positions of the band centres for the combined peak. This is most clearly shown in Fig. 3.6d(i) The positions of the true band centres are shown by the black dots. As the overlap of the peaks decreases, the positions of the true and apparent band centres converge and becomes identical with the peak maxima. For simplicity, therefore, we shall refer to peak height ratios rather than the more correct band size ratios, even though it is not possible to pick out the positions of the peak maxima in the case of poor resolution. However, we shall see later that this has an affect on the measurement of the peak area or peak height measurements required for quantitation in chromatography.

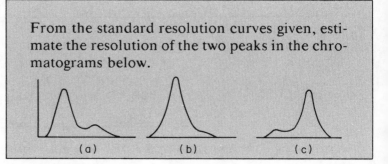

SAQ 3.6d   From the standard resolution curves given, estimate the resolution of the two peaks in the chromatograms below.

(a)   (b)   (c)

Value of R_s Required

The actual value of R_s required will depend on the information needed from the chromatogram. Broadly speaking, this information can be of four types, though of course different types of information may be combined.

These are:

(*i*)   How pure is a sample?

(*ii*) What compounds are present?

(*iii*) How much of each compound is present?

(*iv*) In what purity can these compounds be recovered?

How pure is a sample?

Usually, following a synthesis, the chemist requires to know whether the sample he has prepared is pure or not, or he may have brought-in a 'pure' compound and wants to check its purity. Running a chromatogram of the sample may confirm the purity by giving a single sharp peak. On the other hand, it may give a chromatogram which he suspects may not be a single peak but may be two strongly overlapping peaks. In Fig. 3.6d(iii), at $R_s = 0.6$ there is a noticeable 'tail' to the peak. This may represent a case of peak tailing due to the wrong choice of chromatographic conditions or it may be two overlapping peaks with their band centres marked by the dots on the chromatogram. Clearly, to answer this question, the peak separations must be improved. If the R_s is increased to 0.8 and an obvious shoulder appears, then we know that two compounds are present. However, if band tailing was involved, the shape of the band would not be changed by increasing R_s.

What compounds are present?

The simplest way to identify a component in a chromatogram is to compare its retention time with that of a standard substance under the same experimental conditions. Identical retention times are in-dicative of identical substances. To use this method we must be able to measure the retention time of the peak with the necessary degree of accuracy.

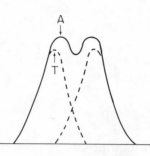

Fig. 3.6e. *Apparent and true positions of band centres in overlapping peaks*

In Fig. 3.6e, the arrows (A and T) indicate the apparent (A) and the true (T) positions of the band centres. Insufficient separation displaces the apparent positions of the overlapping band centres towards each other and reduces the apparent resolution. Thus, the retention times of two overlapping peaks are closer together than the retention times of the two components if injected singly. The standard resolution curves (Fig. 3.6d) indicate the true position of the band centres with a black dot. The horizontal displacement of the apparent band centre from the true band centre allows an estimate of the degree of error in measuring t_R, and the value of R_s when these two points coincide is the value required to give an accurate retention time. In practice a value of R_s of at least 0.8 is required for accurate measurement of t_R.

How much of each component is present?

The usual method of quantifying the chromatogram is by measuring the peak height or peak area. This assumes a proportionality between height or area and concentration. This proportionality is determined by injecting known concentrations of the sample and measuring the peak height or peak area. A calibration graph, ie a plot of peak area (or height) *vs* concentration is then constructed. In dealing with overlapping peaks, the area assigned to each component must be defined. This is often done by dropping a perpendicular from the valley between the two peaks to the base line and assigning the two areas as shown in Fig. 3.6f.

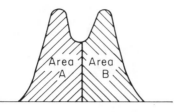

Fig. 3.6f. *Allocation of peak areas for overlapping peaks*

Other ways of assigning the areas will be described later in Section 4.2. An alternative to measuring peak areas is to measure peak heights. Whichever method is used, it is important to know what resolution is required for the accurate quantitation of the peak area or height. Using peak height measurements we have already seen that overlapping peaks shift the position of the apparent peak maximum. In Fig. 3.6d(ii) we see that an $R_s = 1.0$ is required for accurate peak height measurement. Values of $R_s < 1$ would lead to an overestimate of the peak height since the true position of the band centre as indicated by the black dot is to the left of the apparent band centre and at a lower height (see eg $R_s = 0.4$). Using an R_s value of 1.0, the relative concentrations of the two components can vary by a factor of 1000 without there being a significant (ie $> 3\%$) change in the band heights. The effects of poor resolution on peak area measurement are less easy to assess. As the relative concentration of the minor component decreases, the accuracy of the area measurement also decreases. The error in the area measurement of the larger peak is now greater than 1%. If the peak height ratio of two overlapping peaks is x ($x > 1$) the error in the measurement of the peak areas is related by:

$$\% \text{ error in major band} \ = \ - (1/x)\% \text{ error in minor band.}$$

If 100% error represents the case of no error in the minor band then Fig. 3.3g shows the error in quantitation of the minor peak with an R_s value of 1.0 for different band area ratios.

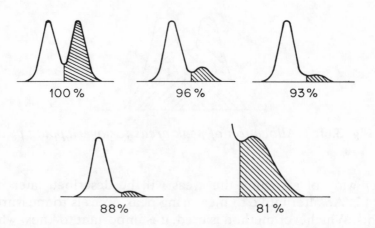

Fig. 3.6g. *Error in quantitation by band area (100% equals no error in minor band)*

For a ratio of 16 : 1, the area error of the minor band is -12% (100–88), so that the error for the large peak is $(+\ 12/16)\% = 0.8\%$. For an $R_s = 1.0$, and for an accuracy of less than 3%, the relative concentrations of the two components can vary by no more than a factor of 10. This means that the resolution must be better for quantitation by peak area measurement than by peak height measurement. The error introduced by poor resolution in quantitation by peak area can be estimated by the ratio of the valley height (h_v) to the height of the minor peak (h_2).

h_v/h_2	Relative error in area of minor peak(%)
0.25	-1
0.40	-2
0.60	-5
0.75	-10

Thus, the accuracy of the quantitation decreases rapidly as the valley between the two peaks disappears.

What is the purity of recovered components?

In a preparative separation, the factors of concern are the amounts of sample recoverable and the purity of those samples. The standard resolution curves (Fig. 3.6d) can again be used since on these curves (at the higher R_s values) are *numbered arrows* between the band centres. These arrows represent the point at which a cut should be taken to obtain two fractions of *equal purity*, the number giving the purity eg in Fig. 3.6d(ii) at $R_s = 0.8$ both fractions can be obtained in 96% purity if the cut is made at the point indicated. If a higher purity is required, we can either improve the resolution to, say, 1.25 and obtain 99.5% purity, or we could reject the region of peak overlap so that region of the peak with 100% purity (or indeed some intermediate purity) was collected. This of course reduces the amount of material recoverable but may be easier and quicker than improving resolution.

The recovery of the sample can be estimated from the standard resolution curves in combination with Fig. 3.6h as follows.

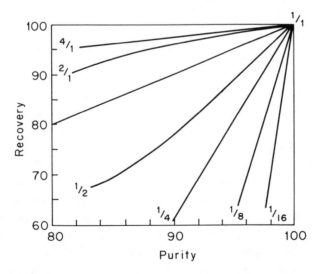

Fig. 3.6h *Estimating the recovery of each band (compound) using the equal-purity cut-point*

First determine the peak height ratios and the value of R_s.

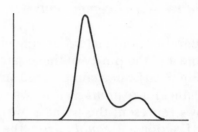

Fig. 3.6i.

In Fig. 3.6i, the peak height ratios are 4 : 1 for the major peak and 1 : 4 for the minor peak. Comparison with Fig. 3.6d shows that R_S = 1.0 and the cut-point gives equal purity of 98%. The percentage recovery can now be estimated from Fig. 3.6h by looking at the 4/1 curve and reading off the recovery corresponding to 98% purity. This gives a recovery of the major peak of 98%. Similarly for the minor peak using the 1/4 curve, the recovery is approximately 94%. Thus, the major peak is 98% recovered with 98% purity and the minor peak is 94% recovered with 98% purity.

In deciding on what resolution is required for a given purpose it is best to err on the side of over resolution. This allows for errors in the estimate of the required change in R_S, difficulties experienced with the practical conditions needed to give the required R_S and any degradation of the column performance with time which leads to a gradual decrease in R_S. Estimations of fraction purity and equal-purity cut-points assume that the detector response is the same for both compounds. In gas chromatography using a flame ionisation detector, this is often the case, but in liquid chromatography, be-cause of the response characteristics of the detectors used, it is rarely the case. In the absence of information on detector response, there is no better way of estimating impurity and cut-point however. Al-though it is desirable to err on the side of caution, over resolution of peaks should also be avoided as it merely increases the analysis time.

So far we have dealt only with pairs of peaks and indeed resolution can be measured only in this way. However, in a real chromatogram, there may be many peaks and the value of the resolution will depend on the choice of such peaks. It may be that not all the peaks are of interest. Perhaps we are interested only in a drug and a metabolite. In this case, we can simply disregard all other peaks and adjust the resolution to the required value for these two peaks. If we are interested in several peaks, the procedure is to select the *least* well resolved pair of peaks, adjust the chromatographic conditions to give the required resolution for these two peaks and assume that the resolution of all the other peaks is satisfactory, since it will be at least as good as, if not better than, our chosen pair of solutes. The overall resolution for the chromatogram can then be quoted as this minimum value.

SAQ 3.6e

For the chromatogram given below

(a) determine the R_s value for the chromatogram and

(b) calculate the R_s value for peaks 4 and 5 and the percent recovery at the equal-purity cutpoint.

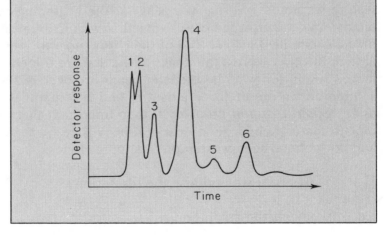

SAQ 3.6e

3.6.3. Optimisation Equation

In our 'typical chromatogram' at the beginning of Part 3 the final feature we noted was that the peaks at the beginning of the chromatogram were crowded closely together with many peaks overlapping and the peaks at the end of the chromatogram were over separated. In other words, the peaks at the start of the chromatogram lacked resolution whereas the later peaks are over-resolved. This is a common feature of chromatograms and is sometimes referred to as the *general elution problem*. It is so important that in any book on chromatography a good deal of time will be spent on this topic, and books have been written on this topic alone.

Π   How could we optimise the resolution?

Eq. 3.19 defines the resolution as

$$R_s = \frac{t_{R_2} - t_{R_1}}{\frac{1}{2}(W_1 + W_2)}$$

so that it is given by the separation of the band centres ($t_{R_2} - t_{R_1}$) and by the peak widths, which are a function of the column efficiency.

Intuitively then, we might guess that if we use a longer column we should improve resolution since a longer column will contain more theoretical plates (ie will be more efficient). The longer the column the longer the migration distance of the solutes and the greater the difference in $t_{R_2} - t_{R_1}$. But remember that the longer a peak takes to pass through the column the broader it becomes so that although Δt_R increases so does $W_1 + W_2$.

∏ Does Δt_R increase more rapidly than $(W_1 + W_2)$ with an increase in column length?

It can be shown that Δt_R is proportional to the column length (L) and $(W_1 + W_2)$ is proportional to L, so that an increase in column length will always produce an increase in resolution. However, there are practical problems to this approach. Eq. 2.16 shows that the retention time is directly proportional to the column length and inversely proportional to the mobile phase flow rate, so that increasing the column length will increase the retention time unless the flow rate is also increased in the same ratio. With a longer column we must have more pressure to drive the mobile phase through, so that column length is ultimately governed by the available pressure. The generation of high pressures involves more robust tubing and connecting joints (and more wear on pump seals etc in liquid chromatography). This, together with the problems of packing very long columns and housing them in a suitable container (eg oven), limits the length of column in practice. Thus, Eq. 3.19, whilst giving us a value of R_S, is of little value in the optimisation of resolution.

Eq. 3.21 shows that the resolution of two components (1) and (2) can be related to the capacity factor (k'), the relative retention (α) and the number of theoretical plates (N_2) in terms of the second component of the pair by the expression:

$$R_S = \frac{1}{4}\left(\frac{\alpha - 1}{\alpha}\right)\left(\frac{k_2'}{1 + k_2'}\right)N_2^{\frac{1}{2}} \qquad (3.21)$$

It is useful to consider the three terms on the RHS of Eq. 3.21 as independent functions in order to investigate further the effect these terms have on resolution, ie selectivity $\qquad (\alpha - 1)/\alpha$

capacity factor $\qquad k_2'/(1 + k_2')$

column efficiency $\quad N^{\frac{1}{2}}$

The selectivity and capacity factor can be described as thermodynamic factors and the column efficiency as a kinetic factor.

Selectivity α

By assuming that the capacity factor and efficiency terms, are constant, the variation of R_s with α may be obtained (Fig. 3.6j).

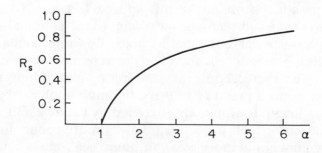

Fig. 3.6j. *Variation of resolution with selectivity*

If $\alpha = 1$, it means that the distribution coefficients (K_1 and K_2) of the two components are the same and there can be no separation. At least partial separation can be achieved in systems with α as low as 1.01.

The main factor governing α is the nature of the stationary phase, although in liquid chromatography the mobile phase composition will also affect it. Temperature also has a small effect in both liquid and gas chromatography.

Capacity Factor k'

By assuming that the selectivity and efficiency terms are constant, the variation of R_s with k' may be shown (Fig. 3.6k).

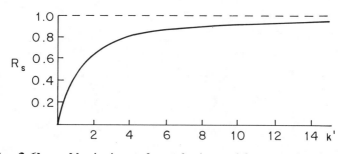

Fig. 3.6k. *Variation of resolution with capacity factor*

If $k'_2 = 0$ then $R_s = 0$ and again there is no separation. Since Eq. 2.9 gives

$$t_R = t_m (1 + k')$$

a substance with $k' = 0$ is eluted at time t_m, ie the hold-up time, and moves with the same velocity as the mobile phase. For low values of k'_2, resolution increases very rapidly with increasing k'_2, though for $k'_2 < 1$ the resolution is too small to be of use. At large values of k'_2 the term

$$\frac{k'_2}{1 + k'_2} \rightarrow 1$$

and further increases in k'_2 will not improve resolution. Thus, the optimum values of k' are in the range $1 \leq k' < 10$. The capacity factor is controlled largely by the mobile phase composition in liquid chromatography.

Column Efficiency N

By assuming that the capacity factor and selectivity terms are constant, we see that the resolution will increase with an increase in N_2. Since the resolution is proportional to $N_2^{\frac{1}{2}}$, and N_2 is proportional to the column length, resolution is also proportional to $L^{\frac{1}{2}}$, ie a two-fold increase in column length (and hence N) will increase R_s by a factor of only 1.4, but, as already noted, the analysis time will be increased by a factor of two.

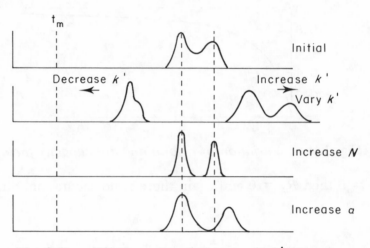

Fig. 3.6l.   *Effect on R_s of changes in k' N and α*

The effect of varying these terms is illustrated in Fig. 3.6l. A change in k' may have a dramatic effect on the resolution. A decrease in k' will further reduce resolution whereas an increase in k' will increase R_s. From Eq. 2.9 we see that an increase in k' will increase the retention time and will lead to a decrease in the band height. However, no other change in separation conditions will give as large a change in R_s with as little effort, and the conditions should first be adjusted so that k' falls in the optimum range of $1 \leq k' < 10$.

An increase in N will produce narrow, higher peaks and an increase in R_s, but the retention times will not be affected. Changes in α will shift the band centre of one peak relative to another and may lead only to a re-ordering of the peaks without the overall resolution of the mixture being affected. However, if α is very close to 1.0 a change in α will produce a large change in resolution, and this approach should be considered for solutes with low α values.

SAQ 3.6f	Describe a general practical strategy for obtaining a satisfactory resolution in terms of k', N and α.

SAQ 3.6f

SAQ 3.6g The figure below shows different resolution problems (*a*) (*b*) and (*c*) requiring different separation strategies. Which of the three parameters k' N and α would you change to produce the desired resolution in each case?

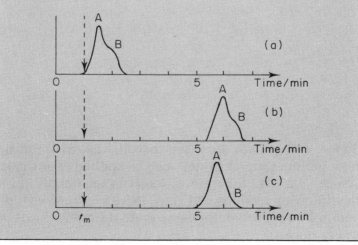

SAQ 3.6g

So far, we have not discussed specific ways of optimising the resolution, but the general strategy can be applied to any chromatographic system. We shall look more closely at the details of optimisation in the following sections and since these are somewhat different in gas and in liquid chromatography these differences will be emphasised where necessary.

3.6.4. Optimisation of Column Efficiency

In Section 3.3 we looked at the processes which lead to the broadening of the chromatographic band and obtained Eq. 3.8 which gives the plate height (H) in terms of various column parameters. The object of optimisation is to make H as small as possible. Fig. 3.5e showed how H varies with the mobile phase velocity $\bar{u}$, and in order to optimise the column efficiency it helps to be able to break down these curves into the general terms A, B, and C of the van Deemter equation (3.15).

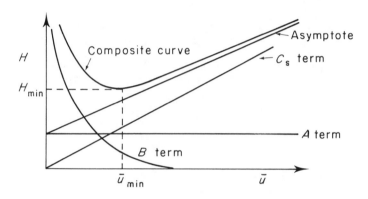

Fig. 3.6m. *Relation between the curve of the Van Deemter equation and the terms A, B and C*

Fig. 3.6m shows the composite curve for H and its three component curves. We see from this that the A term is independent of $\bar{u}$ (some work indicates that this may not be strictly true but we will consider that it is independent of $\bar{u}$), the B term is important at low flow rates, and the C_s term is most important at high flow rates.

The composite curve has the mathematical form of a rectangular hyperbola, so that the values of A, B and C are related to the asymptote to the curve (see Fig. 3.6m). Values for A, B and C may be obtained graphically as follows.

Construct the asymptote to the curve at fast flow rates. This may be done by placing a transparent ruler in the approximate position of the asymptote and adjusting the ruler until the vertical distance between ruler and the curve is inversely proportional to the flow rate. The intercept of the asymptote on the ordinate ($\bar{u} = 0$) is A; B is the constant product of the flow rate and the distance between the asymptote and the curve, and C is the slope of the asymptote.

SAQ 3.6h
The figure below shows the H versus $\bar{u}$ curve obtained on a packed column in gas chromatography. Calculate the values of A, B and C in the van Deemter equation:

$$H = A + B + C\bar{u}$$

[It may be helpful to redraw the curve onto graph paper.]

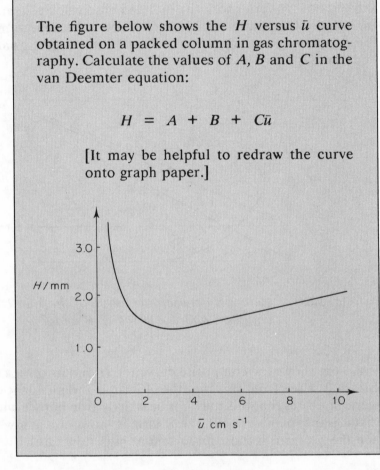

SAQ 3.6h

The A term, $2\lambda d_p$, implies that H can be reduced if the particle diameter d_p is reduced. However, as d_p decreases it becomes more difficult to pack the column so that the packing factor, λ, increases. λ has its smallest value for spherical particles with a narrow distribution of particle sizes. Furthermore, the permeability of the column decreases as d_p decreases so requiring higher pressures to force the mobile phase through the column. In gc this larger pressure drop helps to nullify the improvement in H. For these reasons, the particle diameter used in gc is in the range 100–200 μm whilst in hplc the particles are much smaller, 3–10 μm commonly.

The B term involves the coefficient of diffusion of the solute species in the mobile phase (D_M). This term is important with a gaseous mobile phase since the solute will have a smaller value of D_M in a gas of higher relative molecular mass (greater density). Thus, of the common carrier gases, nitrogen will be superior to those of low relative molecular mass such as hydrogen and helium. However, H versus $\bar{u}$ plots for these three gases show that nitrogen achieves its advantage only at very low flow rates, so that for fast analysis, hydrogen (and helium) are to be preferred to nitrogen.

The mass transfer terms (C) are complex involving diffusion coefficients in the stationary and mobile phases $(D_s$ and $D_m)$ the film thickness (d_f) and the capacity factor k'. In gas chromatography, the most important factor is the film thickness which should be kept to a minimum. In hplc with bonded stationary phases, the concept of film thickness is a difficulty since the orientation of the bonded molecules on the silica substrate may depend on the mobile phase composition, and it certainly depends on the method of preparation. Also, changes in film thickness will alter the k' values so that the two terms are interdependent. Both moving mobile phase mass transfer and stagnant mobile phase mass transfer in liquid chromatography are strongly dependent on the particle diameter, so that small (3–10 μm) particles are used.

The mobile and stationary phases are chosen because of their ability to give the required selectivity and k' values, not because of their diffusion properties. However, when there is a choice between two stationary or mobile phases, then the choice should be made in the light of the van Deemter and similar equations, a low mass transfer term being favoured by high rates of solute diffusion in the stationary and mobile phases.

An elevated temperature will improve mass transfer and hence decrease H, but as we shall see later, in gc temperature is used in a far more dramatic way to control resolution.

SAQ 3.6i	From a consideration of the van Deemter equation predict the affect of the following changes on the value of the plate height H.
	(i) a decrease in particle size
	(ii) the use of a narrow particle size range
	(iii) the use of spherical particles
	(iv) an increase in the density of the mobile phase $\longrightarrow$

SAQ 3.6i
(cont.)

(*v*)   an increase in mobile phase pressure

(*vi*)   an increase in flow rate

(*vii*)   a decrease in temperature

(*viii*) a decrease in the amount of stationary phase

(*ix*)   a decrease in the viscosity of the stationary phase

A more detailed study of the optimisation of the column is outside the scope of this Unit but the general approach we have used can be extended to capillary columns for gc and to hplc columns.

Once the column parameters have been optimised, it is probably easier to deal with gas and liquid chromatography separately as the two techniques differ somewhat due to the effect of the mobile phase.

3.6.5. Optimisation of Resolution in Gas Chromatography

In gas chromatography because the mobile phase plays no part in selectivity it need not be considered in the optimisation of resolution. Eq. 3.22 gives the clue to the approach to be used in gas chromatography. From this equation,

$$\frac{(V_g)_2}{(V_g)_1} = \frac{\gamma_1 \, p_1^o}{\gamma_2 \, p_2^o} \tag{3.22}$$

we see that the ratio of the retention volumes for the components (1) and (2) is a function of the activity coefficients (γ_1 and γ_2) and the saturated vapour pressures (p_1^o) and p_2^o) of the solutes. The vapour pressure is related to the temperature by

$$\ln p^o = \frac{-\Delta H^o}{RT} + \text{constant} \tag{3.23}$$

where ΔH^o is the molar enthalpy of vaporisation. A plot of log V_g versus $1/T$ is linear so that changes in temperature will affect the retention of a solute in a predictable way.

The control of retention by changing the temperature of the column is known as *temperature programming*.

∏   Referring to our typical chromatogram in Fig. 3.6n, and assuming that the chromatogram was obtained at a constant temperature (ie *isothermally*) of 100°C, suggest how you could change the resolution of the peaks in the chromatogram by a change in temperature. Can you divide the chromatogram into regions of differing resolution?

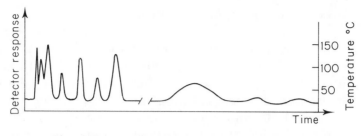

Fig. 3.6n.   *(i) Typical chromatogram*

It will be helpful if we superimpose the temperature profile onto the chromatogram, and since the chromatogram was obtained isothermally this will be represented by a straight line (shown —) in Fig. 3.6n *(ii)*. I hope that you identified three distinct regions *(a)* *(b)* and *(c)* as shown in Fig. 3.6n *(ii)*. In region *(a)*, the components are eluted too rapidly so that the column has not had sufficient time to produce a separation. Region *(b)* shows good resolution between peaks whilst in region *(c)* the retention times are too long and the peaks broad and diffuse.

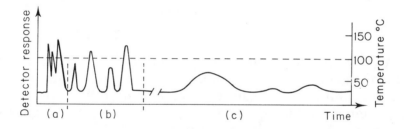

Fig. 3.6n.   *(ii) Typical chromatogram with regions of differing resolution and horizontal isotherm*

Π   What affect would we see on the chromatogram if we re-run the sample at (1) 50 °C and (2) 150 °C?

At 50 °C, the components in region *(a)* would be eluted more slowly and would be better resolved, but those in both the other regions would also be eluted more slowly, so that the peaks in region *(c)* in particular would be even broader. At 150 °C, all the components would be eluted more rapidly, so that although peaks in region *(c)* would be sharper, peaks in regions *(a)* and *(b)* would be less well resolved.

∏   How can we overcome this problem?

By using a *temperature programme*. If we start the run at 50 °C and keep the temperature constant until the first three peaks have eluted (say 3 mins), we can then increase the temperature at a predetermined rate (say 8 °C min^{-1}) to a maximum temperature of 150 °C and then hold the temperature at 150 °C until the final peak has eluted. The final chromatogram should then look like Fig. 3.6n (*iii*).

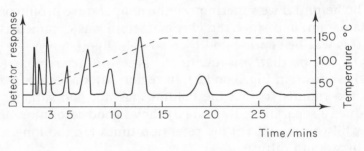

Fig. 3.6n.   *(iii) Typical chromatogram with temperature programming*

In this example, we have used a single ramp programme with a linear temperature rise, but more complex programmes could be used depending on the nature of the mixture to be separated and the limitations of the equipment (eg a multi-ramp programme with convex or concave temperature rises).

The same effect, in theory, could be achieved by using *flow programming* rather than temperature programming. Eq. 2.16 shows that the retention time is inversely proportional to the mobile phase velocity. If the flow rate was programmed in an anologous manner to the temperature, a similar result could be obtained. However, in practice the accurate control of gas flow rates is not easy and because of the compressibility of the gas any change in pressure to change the flow rate involves a time delay that makes reproducibility much more difficult.

If the saturated vapour pressures are very close together, eg for cyclohexane and benzene (boiling points 80.6 and 80.1 °C respectively), then a change in temperature may not produce the desired affect, and another approach must be used.

If the stationary phase is changed, then the ratio of the activity co-efficients (γ_1/γ_2) may also change so that the ratio of the retention volumes may alter to give a better separation. In Section 2.3.3 we saw how the affinity of a solute for the stationary phase could be thought of in terms of polarity, and the adage *like has an affinity for like*. The polarity of the stationary phase may be changed so as to effect the desired separation, but, as already pointed out, in a complex mixture this may cause only a re-ordering of the peaks without an improvement in the overall resolution of the chromatogram. Furthermore, changing the stationary phase means changing the chromatographic column, and this may be a time consuming process, especially if the new column must be conditioned before it gives reproducible results.

3.6.6. Optimisation of Resolution in Liquid Chromatography

In liquid chromatography, retention is controlled mainly by the nature and composition of the mobile phase. Even with diverse sample types, ranging from non-polar to ionic species, the same stationary phase can often be used for the separation. For example, non-polar molecules can be analysed using a non-polar bonded phase with a polar mobile phase, while ionic species can be separated using the same stationary phase but with a mobile phase containing an ion-pairing reagent.

Except in the more specialised modes of chromatography, such as exclusion or affinity chromatography, the stationary phase of choice will usually be one of the bonded phases because of their versatile behaviour and because such columns give high efficiencies. The majority of analyses are done on the non-polar hydrocarbon type, though for some applications polar bonded phases may be preferred.

The choice of chromatographic mode to be exploited will initially be made with regard to the general comments made in Section 1.2.3, but this may have to be changed in the light of experience gained during the early attempts to produce the desired separation. Unlike gas chromatography, where the problem is usually too much retention on the stationary phase, in liquid chromatography the problem is usually too little retention. The initial conditions normally have to be modified to cope with this.

The first step in the optimisation of resolution is to ensure that the column has an acceptably small H value according to Section 3.5.4.

The second step is to optimise the k' values to between 1 and 10. This is achieved by changing the solvent strength as explained in Section 2.3.3. According to whether we are using the normal or the reversed phase mode of liquid chromatography, the retention will be changed by changing the polarity of the mobile phase in the appropriate direction.

In order to simplify this procedure, we need a 'scale of polarity' so that we know which solvents to use if we wish to increase or decrease the polarity. The three most commonly used measures of polarity are the Snyder solvent strength parameter, ε^0, the Hildebrand solubility parameter, δ, the solvent polarity parameter, P'.

The Snyder solvent strength parameter is based on the adsorption energy of the mobile phase on alumina per unit area. However, for silica the values are qualitatively of the same order of magnitude. As measured by ε^0, an increase in ε^0 means an increase in solvent strength and a decrease in k'. The solvent strength parameter makes no attempt to quantify the various interactions which are responsible for the adsorption of the molecule, but measures a global quantity which can be related to polarity. In the same way if we measure the dipole moment of a molecule we do not consider the dipoles of the individual bonds in the molecule but we measure the overall effect.

Both the Hildebrand solubility parameter (δ) and the solvent polarity parameter (P') are obtained by summing the contributions from different intermolecular interactions. The Hildebrand solubility parameter (δ) is given by

$$\delta = \delta_d + \delta_o + \delta_a + \delta_h \qquad (3.24)$$

where

δ_d represent the interactions due to London dispersion forces

δ_o represents the dipole interactions, and

δ_a and δ_h represent the interactions due to proton acceptor (basic) and proton donor (acidic) interactions.

The solvent polarity parameter (P') is based on experimental solubility data where

$$P' = \log(K_g)_{\text{ethanol}} + \log(K_g)_{\text{dioxane}} + \log(K_g)_{\text{nitromethane}} \quad (3.25)$$

where $\log K_g$ is proportional to the free energy of vaporisation of the solvent and is calculated from solubility data for ethanol, dioxan and nitromethane in the solvent.

$\log(K_g)_{\text{ethanol}}$ reflects the proton acceptor properties, $\log(K_g)_{\text{dioxane}}$ the proton donor properties and $\log(K_g)_{\text{nitromethane}}$ the dipole interactions. In Fig. 3.6o. we list the dipole moments, and the ε^o, P' and δ values, for some common solvents in order of increasing eluent strength according to their ϵ^o values. This series is known as an *eluotropic series*. According to the ε^o values, as we descend the solvents in the table, the eluent strength will increase which, in a normal phase system, will give a *decrease* in the k' value, eg replacing heptane by chloroform will decrease k'.

Solvent	ε^0	δ	P'	Dipole Moment/D
n-pentane	0.00	7.1	0.0	0.0
n-heptane	0.01	7.4	0.0	0.0
cyclohexane	0.04	8.2	0.0	0.0
trichloromethane	0.40	9.1	4.4	1.87
dichloromethane	0.42	9.6	3.4	1.6
tetrahydrofuran	0.45	9.1	4.2	–
1,4-dioxan	0.56	9.8	4.8	0
ethyl ethanoate	0.58	–	4.3	1.8
acetonitrile	0.65	11.8	6.2	3.5
propan-2-ol	0.82	10.2	4.3	1.7
methanol	0.95	12.9	6.6	1.7
water	1.00	–	10.2	1.85

Fig. 3.6o.   *An Eluotropic series of solvents*

3636

SAQ 3.6j

In a reverse phase system (using, say, a non-polar stationary phase, what would be the effect on the k' values of changing the mobile phase from methanol to acetonitrile?

SAQ 3.6j

Looking at the dipole moments for cyclohexane and dioxan, we see that they are both zero. However, they differ very much in their ε^0, δ and P' values, indicating the different nature of the two molecules

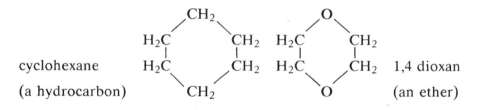

cyclohexane

(a hydrocarbon)

1,4 dioxan

(an ether)

Although trichloromethane and ethyl ethanoate have similar P' values, their positions in the eluotropic series, and hence solvent strengths, are quite different and they would give different k' values. On the other hand, trichloromethane and dichloromethane have similar ε^0 values but different P' values, reflecting the difference in chlorine content.

The importance of these scales will become apparent when we look at the optimisation of the final term, the relative retention α, which is a measure of selectivity. The solvents may be collected together into *Solvent Selectively Groups*, which have similar proton acceptor/proton donor and dipole properties. One such set of groupings, produces eight groups ranging from Group I, which consists of pure proton acceptors such as ethers and amines, through group II, donor/acceptors such as alcohols, to Group VIII consisting of pure donors such as trichloromethane.

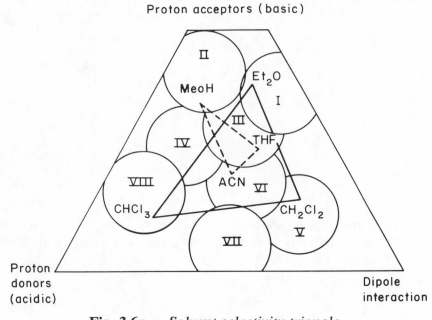

Fig. 3.6p. *Solvent selectivity triangle*

These eight groups can be represented by their position in a triangle (Fig. 3.6p) the corners of which indicate their relative proton acceptor, proton donor and dipole interactions.

To optimise the mobile phase in terms of a mixture of three solvents, it is first necessary to choose them from three different solvent groups. The three chosen solvents should be as far apart from each other in the triangle as possible, bearing in mind that they must be miscible with one other in the proportions in which they are to be used.

Thus, in a normal phase system, trichloromethane (Group VIII) dichloromethane (V), diethyl ether (I) would form a suitable selectivity triangle (−). The optimum proportions of the three solvents can be determined by a series of experiments. Although this can be a tedious process, automated, computerised procedures are available to carry out the optimisation. A fourth solvent, which is neutral in terms of the solvent selectivity groups, is added to adjust the overall solvent strength so as to maintain the desired k' values. Thus, a suitable solvent system for a normal phase separation would be:

$$CHCl_3/CH_2Cl_2/Et_2O/n\text{-hexane}$$

For a reverse phase system, the solvent selectivity triangle (— — —) could be methanol (II), tetrahydrofuran (III) and acetonitrile (VI) which, with water to adjust the overall solvent strength, gives:

$$MeOH/THF/ACN/H_2O$$

The use of ternary (four component) mobile phase systems is not usually necessary to produce the desired separation, and probably represents the ultimate in mobile phase optimisation, but as more complex mixtures and difficult matrices are subject to analysis, such systems will be used more and more. The present literature of liquid chromatography shows that simple binary mixtures of solvents are more often used eg (water/methanol or water/acetonitrile etc) and that the *General Elution Problem* can often be overcome by the simple techniques of gradient elution as described in Section (2.3.2 (2)). Using Fig. 2.3c as the basis for discussion, we can see that in normal phase analysis, increasing the polarity of the mobile phase will lead to a *decrease* in k' values, whilst in the reverse phase mode, an increase in mobile phase polarity will lead to an *increase* in the k' values. Thus, gradient elution allows us to change the solvent strength so that, even in the analysis of a complex mixture, the k' values of all the components can be made to fall within the optimum range.

An alternative to gradient elution (sometimes called *solvent programming*) is that of *flow programming*. We have seen that changes in the flow rate of the mobile phase will change the retention times, but for practical reasons this technique is not suitable for gas chromatography. It can however be used in liquid chromatography where flow rates can be changed easily and reproducibly to give similar results to temperature programming in gas chromatography.

3.6.7. Time Optimisation of a Separation

With increasing pressure on the analyst to achieve a higher through-put of samples, the optimisation of the time for an analysis may become an important consideration. The time for an analysis is approximately equal to the retention time of the last component and can be written:

$$t_R = N(1 + k')(H/\bar{u})$$

Fig. 3.6q shows the variation of $(H/\bar{u})$ and hence t_R with $\bar{u}$.

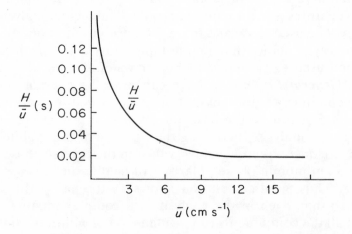

Fig. 3.6q.   *Variation of $(H/\bar{u})$ with $\bar{u}$*

Initially, there is a sharp fall in $H/\bar{u}$ (and t_R) with an increase in $\bar{u}$, but then an increase in $\bar{u}$ produces little change in $H/\bar{u}$. The greatest change in the slope of $H/\bar{u}$ usually lies in the region 2–5 cm s^{-1}, and columns are normally operated within this range of flow rate.

Using the effective plate number,

$$N_{eff} = \left(\frac{k'}{1 + k'}\right)^2 N \tag{3.7}$$

The number of effective plates per second is given by

$$\frac{N_{eff}}{t_R} = \frac{(k')^2}{(1 + k')^3} \, (\bar{u}/H) \tag{3.26}$$

A plot of $(k')^2/(1 + k')^3$ *versus* k' (Fig. 3.6r) shows a maximum at $k' = 2$, so that if we assume H is independent of k', the function $(k')^2/(1 + k')^3$ is directly related to N_{eff}/t_R. Values of k' higher than 2 show little loss in the speed of analysis, but k' values much less than 1.5 should be avoided for fast analysis.

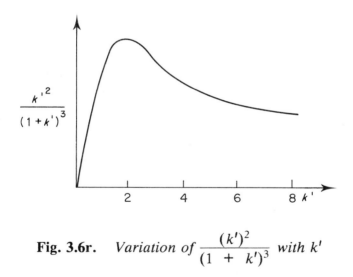

Fig. 3.6r.   *Variation of* $\dfrac{(k')^2}{(1 + k')^3}$ *with* k'

Learning Objectives

You should now be able to:

- describe the main features of a chromatogram;

- understand the relation between peak shape and the sorption isotherm;

- be able to measure column efficiency and understand the factors which affect efficiency;

- explain the processes which lead to the broadening of a solute band as it passes through the column;

- understand the term selectivity;

- define the resolution of a column, and understand resolution in terms of selectivity and column efficiency;

- describe how the resolution of a mixture of compounds can be optimised in gas and in liquid chromatography.

4. Qualitative and Quantitative Analysis by Chromatography

It is easy to underestimate the importance of chromatography in chemical analysis; over the past forty years chromatographic techniques have been developed to separate and analyse a vast range of organic and inorganic compounds. They are amongst the most widely used of analytical techniques.

What types of analysis can chromatography perform?

> *Qualitative Analysis* – chromatography can indicate the presence or absence of compounds, elements or ions in a sample. For example, liquid chromatography is used widely to detect the presence of drugs in samples of serum.

> *Quantitative Analysis* – chromatography can also provide information regarding the quantitative chemical composition of a mixture, for example the proportion of an active ingredient in a drug sample.

What information can *not* be provided by chromatography?

Chromatography will not provide an elemental analysis, ie it will not determine the elemental composition of a sample. Because the technique provides comparative data between compounds it acts as the basis for both qualitative and quantitative analysis.

We will start this Part of the Unit by investigating aspects of qualitative analysis.

4.1. QUALITATIVE ANALYSIS BY CHROMATOGRAPHY

4.1.1. Introduction

Let us begin by considering how we can determine the identity of materials separated by various chromatographic techniques. As with most analytical techniques, the accuracy and precision with which an analysis can be performed, depends largely on a defined procedure. Although procedural details will differ between thin-layer, gas and liquid chromatography, many of the principles involved are identical.

4.1.2. Qualitative Analysis by Thin-layer Chromatography (tlc)

In Sections 1 and 2 you will have discovered that components separated by the process of chromatography can be characterised by their degree of affinity for the stationary phase. This degree of affinity is measured in a variety of ways, depending on the actual chromatographic technique used to perform the analysis.

∏   As an initial exercise to remind you of the terminology used, you are asked to review the terms indicated below. Identify the chromatographic technique to which each term applies and give a brief definition of each term. If you are unsure of any details then refer to the appropriate pages of Section 2.

Term	Chromatographic Technique	Definition
R_f		
R_{st}		
t_R		
V_R		

You should have recognised that the retardation factor terms R_f, and R_{st} are used in plane chromatographic techniques such as paper and thin layer chromatography.

They are respectively defined as:

$$R_f = \frac{\text{distance moved by solute}}{\text{distance moved by solvent front}}$$

$$R_{st} = \frac{\text{distance moved by solute}}{\text{distance moved by reference standard}}$$

The terms t_R, *Retention time* and V_R, *Retention volume* are used in column work. Retention time is defined as the time that elapses between injection of a sample and elution to maximum concentration. Retention volume as retention time multiplied by the flow rate.

You may check your definitions with those given in Part 2.

Each of the above parameters is used to describe the behaviour of a substance under a certain set of chromatographic conditions – a fact that allows us to use them in qualitative analysis.

In the first part of this Section we will investigate how to perform qualitative analysis using paper or thin-layer chromatography. Fig. 4.1a shows the chromatogram obtained after separation of a mixture of organic dyes. Standards have been run alongside to facilitate a comparison.

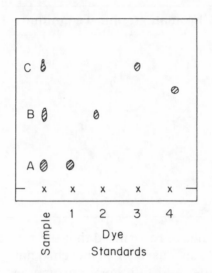

Fig. 4.1a. *Comparison of a chromatographic separation of a sample mixture and a series of dye standards*

∏   The relative positions of spots A, B and C, and hence their R_f values, correspond with the positions of certain standards. Does this confirm the identities of components A, B and C?

The answer is no. Although the positions of the components indicate possible identities, it is also possible for more than one substance to have the same R_f value when chromatographed under a particular set of conditions. If on the other hand, we knew that the sample contained only some or all of the standards used, then this would enable a positive identification of components A, B and C as being standards 1, 2 and 3 respectively to be made. It would also indicate that standard 4 was absent from the mixture. However, in many cases other techniques must be used to confirm the identity of a particular component.

∏   Consider the above example of the separation of some organic dyes using thin-layer chromatography.

Suggest two alternative techniques that could be used, subsequent to separation, to identify component A of the mixture as being identical to standard 1.

You may have recognised that spectrometric techniques such as infrared, ultraviolet and nuclear magnetic resonance spectrometry are all applicable. They are commonly used to identify and elucidate chemical structure. Wet-chemical techniques are generally inapplicable to substances separated by chromatography due to the very small amounts of material available to perform an analysis of this type. Many samples analysed by tlc are present in only μg quantities.

So we have a range of spectrometric analytical techniques available, but practically, due to the difficulty of extracting sufficient quantities of the separated components from the plate, and frequently lack of available access to other instrumentation, we must seek alternative methods of confirmation.

Given these difficulties, it is more practical to compare the appropriate R values of the components of mixtures.

To summarise, what initially seems a simple technique of qualitative analysis, can result in being a quite complex problem if the exact nature of the mixture being analysed is not known.

∏ How can the chromatographer adapt his procedure so that analysis of a chromatogram can give a more reliable identification of a particular component, possibly without resorting to the use of additional analytical techniques?

In the technique of thin-layer chromatography, the reliability of a qualitative determination may be improved by running the same samples under a different set of conditions. Changing the mobile phase or altering the nature of the stationary phase will effect a change in the retention characteristics. The R_f values will thus alter. Fig. 4.1b shows a chromatogram of the same sample as Fig. 4.1a but with a different mobile phase. Notice that the position of component A still coincides with that of standard 1. This fact is further evidence that the identity of component A is standard 1. However, there must still be some reservations as to the precise identity of A, despite this check. It is only by using additional analytical techniques, such as ir, uv and nmr spectrometry, that final confirmation can be made.

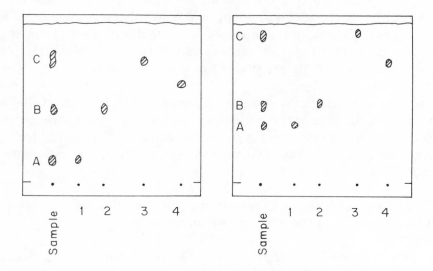

Fig. 4.1b.   *Comparison of sample and standards*
chromatographed under different conditions

Thus, thin-layer chromatography can be used as an aid to the identification of a component by allowing a comparison of the appropriate R values obtained from a number of analyses performed under
different conditions. But absolute identification is possible only if
additional techniques are used.

Nevertheless, in practice this is often not feasible for one reason or
another and reliance is often placed on a comparison of R values.

SAQ 4.1a

A laboratory has been requested to analyse a
mixture known to contain three aromatic hydrocarbons, the identities of which are suggested to
be anthracene, pyrene and phenanthrene.

$\longrightarrow$

SAQ 4.1a (cont.)

The laboratory decides to perform a thin-layer analysis against standards of anthracene, pyrene and phenanthrene respectively that the laboratory has in stock. The resulting chromatogram shows that the three components have R_f values that correspond to the R_f values of the three standards. On this basis the laboratory reports that the identity of the mixture is confirmed.

(*i*) Do you consider their report to be valid?

(*ii*) State the reasons for your conclusion in (*i*)

(*ii*) Suggest alternative means of confirming their result.

4.1.3. Qualitative Analysis by Column Chromatography

Let us now extend our investigation to column chromatographic work. Nowadays, column chromatographic techniques such as gas and liquid chromatography tend to be preferred to thin-layer chromatography because they are more easily automated and the operating conditions such as temperature, flow rates of mobile phase and the nature of the stationary phase can be more easily reproduced and readily altered.

Retardation factors, such as R_f are inapplicable to column systems because there is no equivalent of a solvent front. Hence, separation parameters such as retention time and retention volume are used as a basis for identification. How comparable are the techniques of thin-layer and column chromatography with respect to qualitative analysis?

Tlc has an advantage since it allows the simultaneous analysis of a number of samples. This is not possible in a column, because samples are injected consecutively. How then can the separation of samples and standards be compared?

∏   Fig. 4.1c shows a gas chromatogram of straight chain alcohols. Suggest how peak B could be shown to be ethanol, without resorting to chemical analysis.

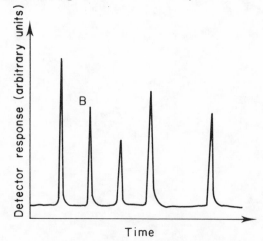

Fig. 4.1c. *Gas chromatogram of linear chain alcohols*

In this particular example, ethanol can be identified because the sample mixture is stated to be straight chain alcohols.

We will deal with each possible response in turn.

Direct comparison can be made between the retention time of component B and that of a sample of pure ethanol injected under identical conditions. This is a quick and valid method, often used. However, the reproducibility of injection is not always good and this can lead to errors in identification.

A better method is to add an additional quantity of ethanol to the sample, after the initial injection, and re-inject this sample. Such a run is shown in Fig. 4.1d. This technique is known as *spiking*. If you compare the chromatograms they appear to be identical, except that the peak marked B has a greater peak area in Fig. 4.1d. This is because an additional volume of ethanol has been added to the injected sample. This technique will positively identify peak B as being due to ethanol, because the method has ensured identical conditions in running both sample and standard.

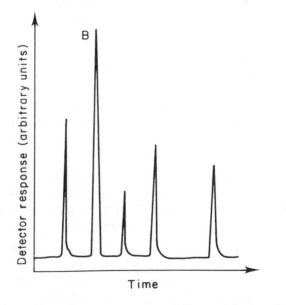

Fig. 4.1d. *Gas chromatogram of linear chain alcohols spiked with ethanol*

The technique of spiking is applicable provided the sample is known to contain certain components; appropriate standards can then be added to the sample to enable each peak to be identified. The technique is *not* applicable to samples in which the components are totally unknown. In this situation, additional analytical or comparative techniques will be required.

Earlier in this Section, you discovered that chromatography can be used in association with other analytical techniques. Gas chromatography (gc) is particularly appropriate for linking to a mass spectrometer. This is because the column effluent is already in the gas phase and can be transferred directly to the inlet port of a mass spectrometer. Spectrometric techniques, in general, have significant advantages for the identification of unknowns. The technique of gc-mass spectrometry has become a particularly powerful and widely used means of identification.

4.1.4. Limitations of Qualitative Analysis

At this stage of investigating qualitative analysis by chromatography, we have started to identify some of the inherent limitations which are in part due to the reliance of the analysis on a comparison between the behaviours of two or more substances. We have listed these limitations as follows:

1. If a sample containing *unknown* compounds is analysed, it will be virtually impossible to identify the components without resorting to analytical techniques other than chromatography.

2. For samples containing *known* components, the identifications can be made by a comparison of retention parameters, *provided* that samples and standards are run under identical experimental conditions and assuming a reproducible spotting or injection technique.

3. The reliability of identifying a particular component will increase with the number of runs made under different sets of conditions.

4.1.5. Comparison of Retention Data

Let us turn our attention to investigate a major problem that laboratories may encounter when they wish to compare their results with those obtained by other laboratories. Consider the following:

Two commercial laboratories perform similar analyses by gas chromatography for the identification of drugs in sera samples using different chromatographs, but run with identical columns under similar operating conditions.

Π   Do you think these laboratories would be justified in comparing their results with each other on a simple retention time basis? If not, could you suggest the reasons why the results may not be comparable.

Ideally, both laboratories would like to obtain results that were directly comparable. But you should be aware that the comparison of retention times may be invalid even on the same instrument, if operating conditions vary between runs. In fact, even columns manufactured from the same batch of stationary phase may have characteristics that differ sufficiently to make direct comparisons of retention times impossible.

To avoid this major problem the following parameters have been proposed to facilitate the comparison of retention data:

Relative retention

Retention indices

With clearly defined chromatographic columns and operational procedures, and isothermal (constant temperature) conditions, sufficiently reproducible data can be collected and reliably compared using these parameters.

Relative Retention

The use of relative retention will eliminate the effect of some variables on a separation, with the exception of the nature of the stationary and mobile phases and the column temperature. The procedure is to compare the retention time or retention volume of an unknown with the retention time or volume of a single standard or reference compound that is chromatographed simultaneously. Because of its dependence on temperature, the use of relative retention in characterisation is applicable to isothermal runs; it cannot be used for temperature programming work.

∏   Can you recall what is meant by the term *temperature programming?*

Temperature programming is a procedure used in gas chromatography, in which the column temperature is raised during a separation to speed up the elution of higher boiling point solutes from a column, Section 3.6.5.

The relative retention, of a substance may be defined as follows:

$$\alpha = \frac{t - t_m}{t_{\mathrm{ref}} - t_m} = \frac{V - V_m}{V_{\mathrm{ref}} - V_m} \tag{4.1}$$

where t and t_{ref} are the retention times of the unknown solute and the selected standard respectively, t_m is the retention time of a compound not retained by the column, V and V_{ref} are the retention volumes of the unknown sample and the selected standard respectively, and V_m is the retention volume of an unretained compound.

∏   Let us see how this works in practice.

The following data were obtained using a gas-liquid chromatograph.

Sample component	Retention time/ minutes	Relative retention
Air	0.11	-
A	1.87	-
B	2.57	
C	3.88	
D	5.81	

You may accept t_m to be 0.11 minutes.

Calculate the relative retention of the components B, C and D with respect to component A.

Since
$$t_m = 0.11 \text{ and } t_{ref} = t_A = 1.87,$$

$$t_{ref} - t_m = 1.76 \text{ min}$$

$$t_B = 2.57$$

$$t_B - t_m = 2.57 - 0.11 = 2.46 \text{ min}$$

If you substitute these values in 4.1 the relative retention time given by:

$$\alpha = \frac{2.46}{1.76} = 1.40$$

Thus, the relative retention of B is 1.40.

The other values are calculated using the same method.

Relative retention of C = 2.14

Relative retention of D = 3.24

Congratulations if you calculated all 3 correctly; if not, recalculate to find your error.

In general, the reproducibility of retention data increases with the chemical similarity of the sample to the chosen standard reference compounds. Fig. 4.1e shows a plot of adjusted retention times against carbon number for a series of alkanes, alcohols and esters.

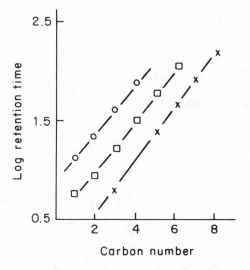

Fig. 4.1e. *Plot of log adjusted retention times against carbon number for n-alkanes ( ), n-alcohols (□) and ethanoates (o)*

However, relative retention times and volumes are not ideal parameters for identification purposes, because they are functions of temperature, flow rate and liquid phase volumes.

Thus, an identification parameter is needed that is independent of these factors. The following method is the best attempt at providing such a parameter.

Retention Indices

A very successful but not perfect solution to the search for an identification parameter is the *retention index.*

Instead of using a single reference material on which relative retention times are based, E. Kovats (Helv. Chem. Acta., 41 1915 (1958)) proposed the use of a homologous series of *n*-alkanes as calibration standards on which to base retention data. Such a method has since become widely accepted for reporting gc data. The basic assumption is that the variation in the retention of hydrocarbons will be reflected in the retention of all other compounds separated under isothermal (constant temperature) conditions. Hence, if the flow rate changes, or the stationary phase volume is reduced, all observed values of the retention volume will change, but the value of the retention index, *I*, will not. The index is not, however, independent of changes in the chemical nature of the stationary phase.

∏   The following data were obtained from a gc column operated under isothermal conditions.

Compound	Retention Time /minutes
air	0.32
n-pentane	2.00
n-heptane	4.79
n-octane	7.40
n-nonane	11.80
n-decane	18.10

(*a*) Plot (*i*) the corrected (net) retention time $(t_R - t_m)$ against carbon number for the series of *n*-alkanes, and (*ii*) the $\log(t_R - t_m)$ against carbon number for the same series.

Assume that $t_m = t_{air}$ for this column.

(*b*) Comment on the shape of the semi-log plot.

(*c*) Does the semi-log plot allow you to predict accurately the retention times of other *n*-alkanes?

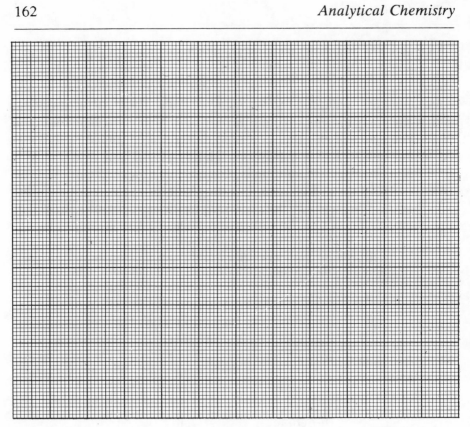

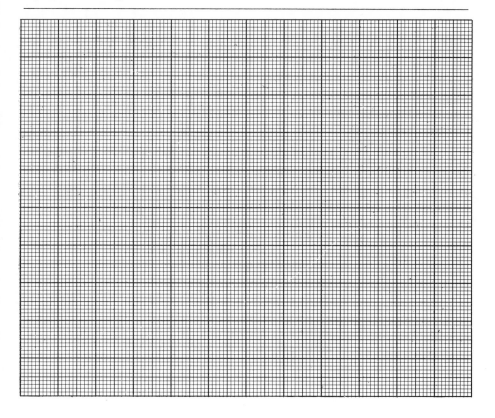

Your plots should appear as shown in Fig. 4.1f and 4.1g.

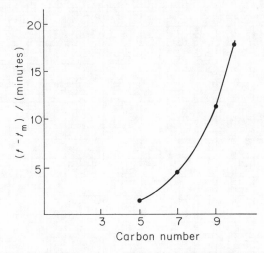

Fig. 4.1f.   *Plot of corrected retention time for a series of samples of different carbon chain length*

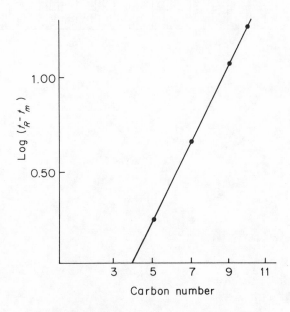

Fig. 4.1g.   *Plot of log (corrected retention time) for a series of different chain length*

The fact that the plot of $\log(t_R - t_m)$ against carbon number is linear should suggest to you that this provides the basis for identifying other *n*-alkanes. Kovats also recognised the fact that if he subsequently chromatographed other compounds under identical conditions, by interpolation from the graph, he could obtain an apparent carbon number for each compound. This empirical number he multiplied by 100 to give a value known as the *Kovats retention index* for that compound.

Π   The next problem will illustrate this point.

The following data were obtained from the same column, operated under identical conditions to that in the previous problem:

Compound	Retention time/ minutes	Kovats Index
Air	0.32	–
A	6.23	
B	7.78	

From Fig. 4.1g, determine the apparent carbon numbers and hence the retention indices for components A and B.

Your answers should be approximately 770 and 870 respectively.

The corrected retention time for solutes A and B respectively is given by $(t_R - t_m)$:

Thus, $t_A - t_m = 5.91$, $\log 5.91 = 0.772$

and $t_B - t_m = 7.46$, $\log 7.46 = 0.873$

Fig. 4.1g shows the relation between apparent carbon number and the corrected retention time. The carbon numbers corresponding to solutes A and B respectively can be read off the graph. This answer times 100 is equivalent to the Kovats Index.

In an analysis where a solute is referenced to two hydrocarbons, one of which elutes before and the other after it, the Kovats Index (I) may be calculated using the formula:

$$I = 100 \times \left[\frac{\log(t_R - t_m) - \log(t_n - t_m) + n}{\log(t_{n+1} - t_m) - \log(t_n - t_m)} + n \right] \quad (4.2)$$

where t_m is the elution time for a substance not retained by the stationary phase, eg air, t is the retention time of the solute, t_n is the retention time of the n-alkane containing n carbon atoms that elutes *before* the solute and t_{n+1} is the retention time of the n-alkane containing $n + 1$ carbon atoms that elutes *after* the solute.

∏   (*a*) Using the data in the previous question, calculate the Kovats indices for samples A and B using eq. 4.2.

(*b*) Compare your values with those obtained graphically.

In this question, sample A is referenced to the two nearest alkanes that elute before and after it, ie n-heptane and n-octane, with retention times of 4.79 and 7.40 minutes respectively.

$$t_m = 0.32 \qquad t_A = 6.23$$

$$t_n = 4.79 \qquad t_{n+1} = 7.40$$

$$n = 7$$

By substituting into 4.2.

$$I = 100 \, \frac{(0.122)}{(0.200)} + 100 \, n$$

Hence, $I = 761$

Isothermal conditions have been stipulated as a requirement for characterisation by relative retention. In a temperature programmed run, the logarithmic relation between retention time and carbon number for *n*-alkanes is not linear, but it can be replaced by an approximately linear relation:

$$I = 100 \times \left[\frac{(t_R - t_m) - (t_n - t_m)}{(t_{n+1} - t_m) - (t_n - t_m)} + n \right] \tag{4.3}$$

In practice, other hydrocarbons can have retention indices related to the Kovats series. However, work has shown (Anal Chem 51 768 (1979)) that when the retention indices of polynuclear aromatic hydrocarbon are recorded using *n*-alkanes as standards, the results show poor comparability. The reliability of retention data increases with the similarity of the sample component to the chosen standard reference compounds.

4.1.6. Qualitative Analysis by Liquid Chromatography

Many of the considerations regarding qualitative analysis that we have discussed so far for gas chromatography are applicable to liquid chromatography. For example, techniques such as spiking, the retention parameters such as relative retention and specific retention volumes are used in both techniques.

However, there is no equivalent to the Kovats Index for liquid chomatography, although there are examples where attempts are being made to provide a system whereby comparability can be made between laboratories on a more reliable basis.

Comparisons in lc are easier since retention is not so dependent on temperature.

SAQ 4.1b	The corrected retention times t_R, in seconds for the following series of compounds were determined on a non-polar (apolar) column. Calculate, using a graphical and non-graphical method, the Kovats Indices for methylbenzene and cyclohexane relative to the series of *n*-alkanes. $\longrightarrow$

SAQ 4.1b
(cont.)

Compound	Corrected Retention Time t_R/s
n-butane	7.1
n-pentane	17.8
n-hexane	38.0
n-heptane	83.0
n-octane	182
methylbenzene	100
cyclohexane	79.0

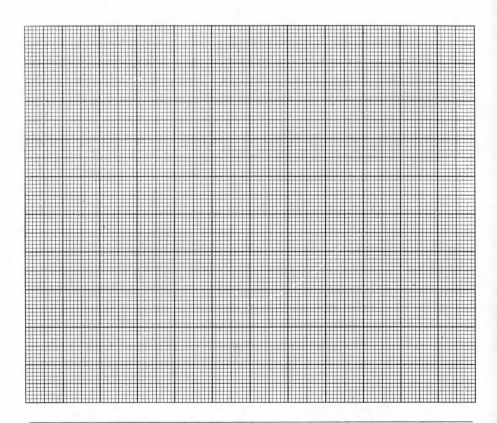

SAQ 4.1b

4.2. QUANTITATIVE ANALYSIS BY CHROMATOGRAPHY

4.2.1. Introduction

It is the purpose of this section to help you realise some of the problems of quantitative analysis from a chromatographer's point of view. In the following sections, we will be investigating quantitative analysis with respect to different chromatographic systems. The Section will include the following:

Listing the requirements that need to be satisfied prior to performing a quantitative analysis, such as determining the response of different detectors, establishing different methods of obtaining quantitative data for each technique and finally, making quantitative calculations from chromatographic data.

General Requirements for Quantitation

Even before a quantitative analysis is undertaken, certain pre-requisites are necessary to ensure a valid analysis of the sample, ie

(*i*) the identity of the component to be analysed must be established.

(*ii*) separation of the specified component must be achieved.

(*iii*) sample preparation must be reproducible.

(*iv*) standards of known purity must be available – accuracy will be directly related to the degree of purity of standards used in the determination.

(*v*) a stationary phase that separates components in a reproducible manner.

Finally, if an internal standard is not used,

(*vi*) there must be a constant flow of mobile phase.

(*vii*) sample application or injection must be reproducible.

The reasons behind these pre-requisites will become apparent when details of specific detectors are discussed.

For the sake of clarity, we shall consider firstly the quantitation of gas chromatography and later extend the discussion to other chromatographic techniques.

In quantitative analysis, we rely on the fact that a component will have some property that can be measured, and that this measurement is related to the amount of material present. For example, many chromatographic systems use a detection system that is designed to record a specific property of the separated components. Full details of these will be given in the appropriate units, but at this stage it is necessary to consider quantitative aspects more generally.

Consider the following question, which deals with detector response and will help you to familiarise yourself with the properties of detection systems:

∏   We have given some examples below of typical types of chromatographic detector. By referring to appropriate analytical textbooks, identify the chromatographic technique to which each example relates and the principle on which it depends. For each detector give one example of the class of compound for which it is best suited. Do not read in depth about each detector, because they will be dealt with fully in other ACOL Units; simply abstract the required information.

Detector	Chromatographic system	Principle of operation	Principal class of compound detected
Flame ionisation			
Thermal conductivity (Hot wire or katharometer)			
Electron capture			
Uv absorption			
Refractometer			

The completed table is reproduced below. You will observe that certain detectors will be selective for certain types of component, while others are universal, meaning that with certain exceptions, the detector will detect all components passing through it.

Detector	Chromatographic system	Principle of operation	Principal class of compound detected
Flame ionisation	gc	Ionisation of solute molecules in a flame	organics
Thermal conductivity (Hot wire or katharometer)	gc	thermal conductivity	any
Electron capture	gc	current reduction by electron capture	compounds containing electro-negative elements
Uv absorption	lc	absorption of electro-magnetic radiation	uv absorbing species
Refractometer	lc	measures change in refractive index	any

In each of the above cases, the output of the detector can be measured as a voltage that will vary with the amount or concentration of the substance detected. This voltage is recorded as a function of time by a chart recorder to produce a chromatogram.

∏   A typical chromatogram for an analysis of alcohols obtained
        using a flame ionisation detector (fid) is shown in Fig. 4.lc.
        Before a quantitative analysis can be performed on this sam-
        ple, the following questions must be answered:

        1.   How does a detector output react to the quantity of the
              material passing through it? (This property of a detector
              is known as its *response*.)

        2.   How *sensitive* is the detector to low concentrations of
              sample?

        3.   Will the response be the same for all materials?

        4.   How can we relate the size of the peak on a chro-
              matogram to the quantity of material to which it cor-
              responds?

Consider these points for a few minutes and note any comments you
wish to make.

1.   Your answer probably included the statement 'the larger the
        amount of sample injected, the greater will be the response of
        the detector'. To an extent this is true. However, the response
        for a particular detector cannot be predicted precisely without
        performing an experiment to measure how the response changes
        with different samples and different concentrations of sample.

2.   This is another factor that can be determined only experimen-
        tally.

3.   This is unlikely, since for a detector that operates on the prin-
        ciple of absorbance, different responses will be obtained from
        samples that absorb radiation to different degrees, ie have dif-
        ferent molar absorptivities.

4.   Again it is difficult to be definitive on this point. There are dif-
        ferent methods of measuring the size of a peak. Some labora-
        tories use peak heights, others measure peak areas manually,

others use electronic or computing integrators that measure peak areas automatically.

It should be evident that unless we can calibrate a particular chromatographic detector/solute combination, we shall not be able to perform a successful quantitative analysis.

Now that we have identified the problems, let us continue and see how they may be solved. We will assume, for the sake of simplicity, that the signal obtained from the detector is dependent only on the concentration of the solute.

4.2.2. Relation Between Detector Response and Sample Concentration

Let us investigate the response of a typical thermal conductivity detector to a particular hydrocarbon n-decane.

∏ Fig. 4.2a shows the output obtained from the detector as a function of time and operating under optimum conditions when analysing known volumes of the hydrocarbon. The detector response is defined in this case as the peak height multiplied by the attenuation of the detector amplifier.

	$\log R$	
Response (R) (arbitrary units)	Volume (V) μl	$\log (V)$
no response	0.003	
9.3	0.01	
89.1	0.10	
890	1.00	
9998	10.00	
26300	31.6	
38000	100	
50100	316	

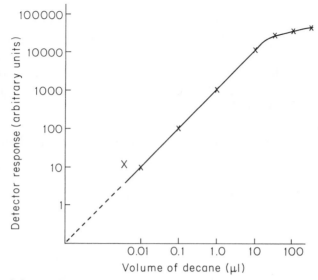

Fig. 4.2a. *Detector response for a typical detector to a series of samples of n-decane*

(*i*) Plot the log of the response against log (volume of *n*-decane).

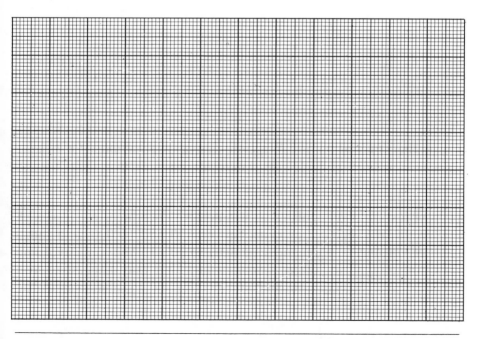

(*ii*) Comment on the shape of the curve and give the range for which the response of the detector is linear.

(*iii*) What would be the consequence of calculating the volume of decane from a detector response value in excess of 50 000?

(*iv*) Suggest why the graph has to be plotted on a log-log basis.

(*i*) Your curve should appear as Fig. 4.2b.

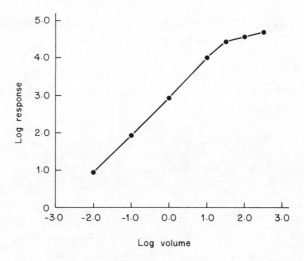

Fig. 4.2b. *Plot of the log Response against log volume of n-decane for a detector*

(*ii*) you should recognise that the detector produces a curve with a linear range of response between 10 and 16 000, ie a range of volume of *n*-decane from 0.01 to 10 μl. Above this volume the plot is no longer linear, a fact that may suggest to you that the detector has become overloaded by this level of the hydrocarbon.

The fact that the response is linear allows one characteristic of the detector to be defined, its *range* of linear response. It is defined as the range of response over a particular range of sample concentration.

(*iii*) at detector response values in excess of 50 000, volumes of *n*-decane calculated would be meaningless, because the volume is outside the range where the log of the response is directly proportional to log of the sample volume.

(*iv*) it would be impossible to plot the data on linear graph paper since the range of the sample volumes and detector responses is far too great. Note the range of linear response of this particular detector.

The concept of linearity will be seen to be important in evaluating a quantitative method.

4.2.3. Sensitivity

Recently, the term *sensitivity* has become synonymous with the term *lower limit of detection* for a particular detector-sample combination. For example Fig. 4.2a indicates that responses below about 0.01 μl cannot be distinguished from noise. At this level, attenuation (amplification) of the signal will be a maximum, but no signal is measureable from the sample passing through the detector. This point is marked as X on the graph and is defined as the lower limit of detection.

The sensitivity of a detector is dependent on the signal-to-noise ratio of the detector system, which in turn is influenced by other properties of the system. It has been defined as a measure of detector response per unit change of amount of sample.

Sensitivity does not indicate the limit of detection since it does not take into account the noise level of the system. From a practical point of view, it is better to quote a limit of detection in terms of the amount of material required to produced a signal 3× greater than the noise level.

4.2.4.   Selectivity

The last problem concerned the response of a particular detector towards *n*-decane. The response is not necessarily the same for other compounds. For example, the response of this detector is dependent on the thermal conductivity of the sample. Unfortunately, we must therefore consider detector responses to each individual component for which a quantitative analysis is required. In the case of some selective detectors, there may be no response at all to certain compounds.

At this stage, you should be able to summarise the important points regarding the detector and its response to components separated by chromatography.

Π   Without referring to the previous section, describe in your own words the meaning of the following terms. If you find difficulty, then re-read the preceding Sections, before answering SAQ 4.2a.

   (*i*)   linearity

   (*ii*) sensitivity

   (*iii*) selectivity

Provided you accept the limitations that the above characteristics present, plots of detector response against sample volume can be used to perform a quantitative analysis. These types of plot are called *calibration curves*, but are usually plotted as linear-linear plots over a more limited range of sample volume.

Different types of detector can show different responses as a function of volume (or concentration) and differ in their responses to individual substances.

As a further example let us study the response of a different type of detector, the electron capture detector (ecd).

SAQ 4.2a

The following values were obtained from an electron capture detector when analysing standards of lindane, a pesticide:

Response (R) (arbitrary units)	log R	Mass of lindane injected (M/ng)	log M
0.56		0.100	
2.01		0.316	
9.77		1.00	
89.1		10.00	
200		100	
224		316	

(i) Plot the log of the response against log M.

(ii) Comment on the shape of the curve, and give the range of linear response of the system and the lower limit of detection.

(iii) Two samples of lindane, extracted from a food sample, give readings of 40.3 and 400 for the detector response. Calculate the mass of lindane extracted from the sample, commenting on your results.

SAQ 4.2a

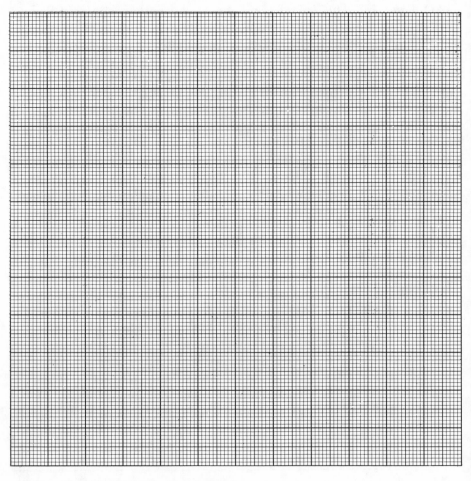

4.2.5. Peak Area Determination

Let us now discover how we may analyse the raw data from a chromatogram.

Detector response is often recorded on a pen recorder, although computing integrators are being increasingly used. It is accepted that the peak size will be directly related to the sample concentration. Our next problem is measuring the peak size, or probably more correctly, obtaining a numerical value that is directly related to the peak size.

The following methods have been used for peak size determinations:

(*i*) measurement of peak height

The measurement of peak height is of course an approximation, but it can be measured with good accuracy and is certainly simpler to perform than measurement of peak area. However, peak height can be more sensitive to variations in experimental conditions such as column temperature and ageing. Fig. 4.2c shows a calibration curve that compares peak height measurement with peak area measurement.

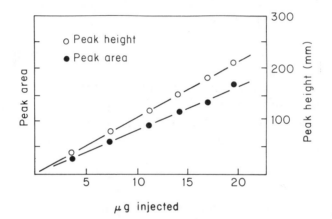

Fig. 4.2c. *Peak height and peak area calibrations for a sample of an antioxidant (J. Chrom. Science, 8, 338 (1970))*

(*ii*) measurement of peak area by triangulation

The area of a Gaussian curve is the product of its maximum height by its width at half the maximum height, the main proviso being that the peak shape should remain independent of quantity of sample injected. (Alternatively, the area is given by the product, 1/2 base × height.) You should note that it is difficult to obtain useful quantitative results from highly asymmetrical peaks and peak height measurement should not be used for peaks that tail badly, unless a computing integrator is available.

The four steps that need to be followed are as follows:

— draw a base line across the peak

— measure the peak height

— draw the parallel to the base at half height

— measure the peak width at half height

It is possible that manual peak area measurement may well be less precise than that of peak height due to the error in measuring the peak width accurately. Precision can be improved by increasing the chart speed of the recorder.

(*iii*) measurement of peak area by cutting out and weighing peaks

This method relies on the homogeneity of the chart paper you are working with and the availability of a very accurate balance. There is the disadvantage that the original chromatogram is destroyed by this method, although photocopying the original will avoid this problem.

(*iv*) measurement of peak area by mechanical methods

An instrument called a *planimeter* may be used to record peak areas by mechanical means and its use can lead to considerably better precision than the previous methods when the peaks have a sufficiently large area. However, the technique has practical difficulties for those inexperienced or untrained in its use and is best avoided.

(*v*) measurement of the peak area by electronic methods

The vast majority of laboratories have electronic or computing methods of integrating the areas of peaks which, depending on the software, can provide by far the most precise method of recording peak areas and much more quickly than manual methods.

Section 4.3 will be devoted to these methods. However, it is still pertinent to review the other techniques that are used, and since practice makes perfect, try the following exercise.

∏ Fig. 4.2d shows a chromatogram with two peaks that are well separated. Compare the areas of the two peaks by four of the above methods. Record in the table the ratio of the size of peak A to the size of Peak B and comment on the methods you have used. (You can photocopy the chromatogram in order to cut and weigh.)

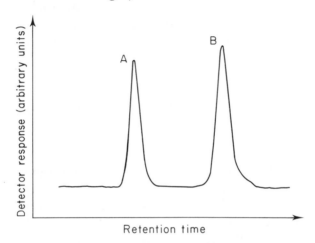

Fig. 4.2d. *Chromatogram of two samples for comparative measurement*

Technique	Size Peak A	Size Peak B	Ratio A/B
Peak height			
Triangulation			
Cutting and weighing			
Planimeter			
Other			

The values we obtained from the chromatogram from which Fig. 4.2d was drawn were:

Technique	Size Peak A	Size Peak B	Ratio A/B
Peak height	66.8 mm	75.0 mm	0.891
Triangulation	394 mm^2	548 mm^2	0.719
Cutting and weighing	0.0392 g	0.05341 g	0.733
Planimeter	0.0460	0.0667	0.684
Other			

It appears that peak height measurement for this application is inappropriate. The other values show reasonable agreement.

4.2.6. Methods of Quantitation of Chromatograms

These methods allow the determination of the composition of a mixture from the peak sizes of the components. In gas chromatography, a principal source of error is sample injection, since it is difficult to introduce quantitatively, small samples of liquids and gases. As a result, the peak areas from seemingly replicate injections may vary considerably from one run to another. Precision can be increased by the use of automatic sampling and injection aids, or by one of the following methods:

(*i*)   internal normalisation

(*ii*)   external standardisation

(*iii*) use of an internal standard

Internal normalisation is a calibration method normally used in gas chromatography. It is based on measuring the area of every peak in the chromatogram, the total area is 'normalised' to 100% and each peak is then reported as a percentage of the total. For certain detector/sample combinations, it can be assumed that response factors for the components are identical; the area of each peak divided by the sum of the areas of all peaks in the chromatogram represents the concentrations of the compounds directly. Obviously, when mixtures are analysed which contain solutes that produce differing responses then calibration and hence additional correction must be applied, ie response factors must be calculated.

∏   The output from a particular detector is recorded during an analysis of a series of straight chain alcohols. The results are recorded below in relative units as peak areas. Calculate the normalised peak areas and hence the mass of each alcohol present. You can assume that the detector output is proportional to mass of the alcohol and independent of the type of alcohol, and that the total mass of alcohol sample injected is 0.0045g.

Alcohol	Peak Size	Normalised Peak size / %	Mass of each alcohol present /g $\times$ 10^4
methanol	1.26		
ethanol	2.42		
propanol	1.88		
butanol	1.06		

The values for all the alcohols are:

Alcohol	Normalised Peak size / %	Mass of each alcohol present /g $\times$ 10^4
methanol	19.0	8.6
ethanol	36.6	16.5
propanol	28.4	12.8
butanol	16.0	7.2

Total area = 6.62

Normalised peak size for methanol $= \dfrac{1.26}{6.62} \times 100 = 19.0\%$

Mass of methanol present = 19% of 0.0045 = 8.6×10^{-4} g

This method is applicable only to analyses where all constituents of the analysed mixture are eluted from the column. It is *not* applicable to mixtures where the constituents are unknowns. The assumptions that are the basis for internal normalisation are only approximate and precision is unlikely to be better than 5% even for mixtures of homologs of similar relative molecular mass.

External Standardisation involves the construction of a calibration plot using standards of known concentration or mass. A fixed volume of each standard is then injected, and the chromatogram analysed in terms of plotting peak size as a function of concentration or mass.

In most systems, calibration plots will be linear and should extrapolate through the origin, although non-linear curves may be obtained. The concentrations of unknowns are read off the curve.

Principal sources of error in this method of calibration arise from changes in the sample injection volume, since syringes are difficult to use reproducibly. Automatic sampling is preferred.

Π   Can you suggest an alternative method of compensating for variations in sample volume injected during a chromatographic analysis?

The technique most commonly used is to add an identical volume or mass of a compound not present in the sample to each sample under investigation and ratio the peak size of the sample component of interest to the peak size of the compound added. This will compensate for variations in injection volume. The method is known as *internal standardisation*. An exact and constant amount of a pure compound known as the *internal standard* is added to a specified volume of the unknown sample and also to several standard mixtures containing known amounts of the constituents of the unknown sample that are to be quantitatively determined. These standards are chromatographed and a plot of the amount of each constituent of interest against the ratio of the peak area of that constituent to that of the internal standard is produced. The sample is then chromatographed and the peak area ratios for each constituent and the internal standard measured. From these ratios, values of the amounts of each constituent in the unknown sample may be calculated.

The technique is used to compensate for variations in sample size and also in analyses where samples may require pre-treatment. There are, however, a number of requirements that the internal standard must meet for it to be successful, ie

The internal standard must:

1. be structurally similar to the compound(s) of interest.

2. be completely resolved from other components of the sample.

3. elute close to the compound(s) of interest (for a multicomponent mixture, it may be necessary to have a series of standards).

4. be stable and of high purity.

Sources of Error in Quantitative Analysis

In addition to errors in the measurement of peak size, there are additional errors introduced by variations in chromatographic parameters that may affect the detector response. For example, the response factor of certain detectors depends not only on the detector parameters, but also on mobile phase flow rate. Any fluctuation in flow rate thus results in an area fluctuation and in practice a variation in response factor.

Nevertheless, chromatography is still a comparatively precise method of quantitative analysis provided that the appropriate procedures are adopted.

SAQ 4.2b

The following data were obtained for a solution of benzene of unknown concentration and a series of standards. An equivalent volume of spectroscopic grade methylbenzene was added as an internal standard to the unknown and the standards. Fig. 4.2e. shows a chromatogram of one of the standards.

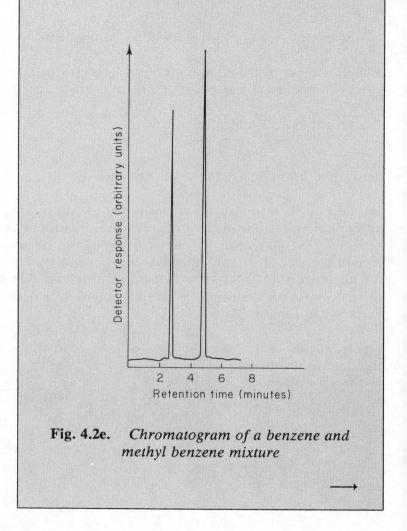

Fig. 4.2e.   *Chromatogram of a benzene and methyl benzene mixture*

→

SAQ 4.2b
(cont.)

(*i*) Does methylbenzene fulfil the criteria necessary to be used as an internal standard for this analysis?

(*ii*) Draw a calibration curve for the benzene standards using the internal standard.

(*iii*) Calculate the concentration of benzene in the unknown.

concentration of benzene /mg dm^{-3}	peak area of benzene	peak area of methylbenzene
2.00	0.504	1.24
4.00	0.969	1.29
6.00	1.45	1.29
8.00	1.93	1.22
10.00	2.47	1.26
unknown	1.34	1.28

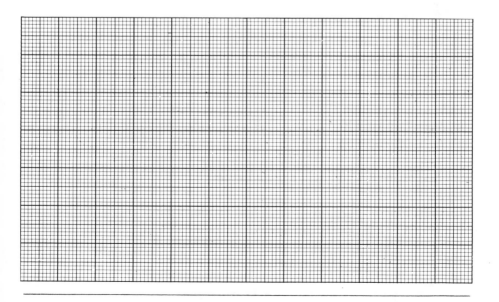

SAQ 4.2b

4.2.7. Quantitative Paper and Thin-layer Chromatography

It is more difficult to perform quantitative analysis in plane chromatography, than in column work. This is due to the greater difficulty in obtaining reproducible conditions and in measuring spot densities.

Standards and samples must be of similar concentrations and applied as spots of similar size. Solvents must be prepared, the chamber brought to equilibrium and locating agents applied in a reproducible manner.

After development, the separated solute(s) can be measured directly on the plate or alternatively removed and measured by analytical methods already discussed.

The following methods are applicable:

1. *Visual comparison* – samples and standard solutions are run on the same plate with a visual comparison made of the spot intensities. The method is unlikely to produce results with a precision better than 10%.

2. *Physical measurement* – scanning spectrometry, such as fluorescence and uv or visible absorption, have been used with success. The spot intensities are plotted graphically and the resulting trace analysed as a conventional chromatogram.

3. *Radioactive measurements* – likely to produce the most accurate results with samples that can be radio-labelled.

4. *Spot removal* – the sample to be analysed is extracted from the stationary phase and analysed using other analytical techniques.

4.3. DATA SYSTEMS FOR CHROMATOGRAPHY

If you have completed all the exercises and SAQ's in the previous Section, you will already realise that the manual interpretation and analysis of chromatograms is time consuming and repetitive. Although such examples may be an aid to understanding, manual analysis of chromatographic data on a regular basis is wasteful in terms of manpower.

In this short Section we will survey the functions of currently available data systems. We will not deal with any aspects of programming because commercial software offers excellent facilities at a price that makes self-programming uneconomic for the majority.

∏ Perhaps you could start by re-analysing the functions of a chromatograph. Itemise the functions and identify those that you consider a computing or data handling facility could perform.

I have listed the following system functions where computers can make an input.

Sample loading – automatic sampling procedures have already been recommended, computer control can be invaluable for laboratories that analyse large numbers of samples, particularly as they can operate around the clock.

Production of a chromatogram – most chromatographs present data as a voltage *versus* time plot on a chart recorder. This produces only hard copy and does not allow manipulation of the data, eg an ability to enhance or decrease peak size, or re-plotting, without re-running the analysis at a different detector attenuation.

Analysis of a chromatogram – procedures such as recognition of peaks, measurement of retention times and relative retention, calculation of relative retentions, data, calibration, standardisation and normalisation are all procedures that could be performed easily by computer.

Reporting of results – many computers have databases that can retain, collate and transfer information. Such a facility can ease report writing and have significant influences on laboratory management.

I hope you agree from the above that there is plenty of scope for using computers in chromatography. Let us continue by looking in a little more depth at some aspects of the above areas.

4.3.1. Data Acquisition

Before a computer can acquire data from a chromatograph, it must be able to 'communicate' with it. The means of communication is usually an interface. For example, many chromatographs still output data to a chart recorder as an analogue (continuously variable) voltage signal. This cannot be read by a digital computer and the signal must be translated into the appropriate digital form. Interfaces provide the facility of analogue to digital conversion. Most systems now operate in a 12 bit mode that can read voltages with a precision of 0.025%.

During data accumulation, a computer will read a signal in a certain time interval (sampling rate). The resolution of the chromatogram held by the computer will depend on this sampling rate.

∏   Fig. 4.3a shows a chromatogram obtained using a sampling rate of two readings per second.

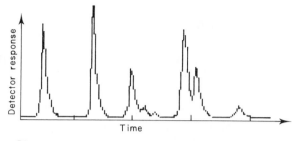

Fig. 4.3a. *Chromatogram recorded at a sampling rate of two readings per second*

(*i*) Can you suggest how the chromatogram shape could be improved?

(*ii*) What implication will this suggestion have on the amount of data that needs to be stored?

The apparent steps in the peaks could be removed by increasing the sampling rate. Fig. 4.3b. shows the same chromatogram, but recorded using a much faster data accumulation rate. Notice that the step intervals are no longer observable and the plot appears as if it were obtained on a chart recorder. However, an increase in the sampling rate by a factor of 5 will also increase the amount of data that needs to be stored by a factor of five.

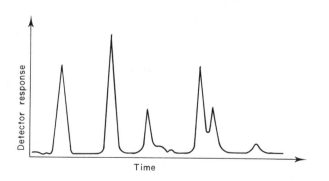

Fig. 4.3b. *Chromatogram recorded at a faster rate than Fig. 4.3a*

For the highest resolution, high sampling rates are required. However, this means that more data has to be stored for subsequent analysis, re-processing or comparison.

∏   Two chromatograms are to be run, one at a sampling rate of one per second and the other at a rate of five per second. If both chromatograms take 30 minutes to run, calculate the minimum memory needed to store each chromatogram. Assume that each data point needs one byte of memory.

Total number of data points for run 1:

30 minutes × 60 × 1 = 1.8 kbytes

Total number of data points for run 2:

30 × 60 × 5 = 9.0 kbytes

For a computer with an available memory store of, say, 50 kbytes, it would not be able to store many higher resolution chromatograms without access to additional disc storage.

4.3.2. Data Storage

Microcomputer systems are now available with memories that can store up to a megabyte or more of information in the working (random access) memory with additional storage in associated disc systems. A single floppy disc can store 400 bytes or more, while a more expensive hard disc can store up to 40 Mbytes. The latter are applicable to the preparation of a library of chromatographic data for subsequent comparison or re-assessment.

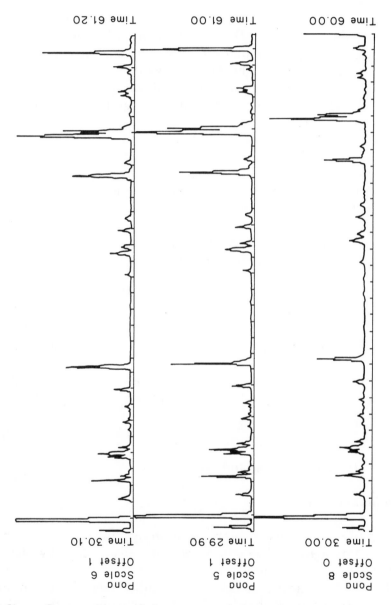

Fig. 4.3c. *Comparison of three sets of data obtained from a disc store*

Fig. 4.3c. shows the ability of a data system to recall three different sets of data and use them comparatively.

Chromatographic systems, without access to computers, record their output onto chart recorders that provide a hard copy. Many computer-based chromatographic systems display the output data directly on a monitor. The major advantage of a computer-based system is that the information recorded can be re-plotted in a number of different ways, without the chromatogram having to be re-run.

We shall list a number of examples:

— the data is re-plotted to increase or decrease the size of peaks in the chromatogram (effectively increasing or decreasing the sensitivity of the original plot)

— the time base may be altered to spread or condense the chromatographic information (useful for comparing chromatograms)

— sections of the chromatogram may be enhanced to analyse shoulders or small peaks

— base line drifts can be corrected

The ability of the system to store and recall without the chromatographer having to refer to hard copy of the chromatographs is highly advantageous.

4.3.3.   Data Processing and Analysis

Qualitative Analysis

In addition to the visual comparisons above, the computer can analyse chromatograms. The computer can 'sense' peaks by identifying the sharp change in gradient of the curve at the peak maximum and shoulders can be detected in a similar manner. By timing the position of the peak maximum and comparing it with an air peak (or one nominated by the operator) the computer can very quickly assign retention times or volumes in relation to the operating conditions established at the start of the analysis. It may wish to use data already established in the computer library and, if necessary, identify the names of the particular compounds it has associated with each peak. However, it is still the operator who must decide on the validity of the results on the basis of the comparative data furnished by the computer.

Quantitative Analysis

Manual measurement of peak size tends to be a very time consuming exercise in quantitative analysis. Computer systems can perform a series of different peak size measurements, compare the results from each method and use the method that gives the most satisfactory results.

Finally, calibration runs can be automatically determined from a series of provided standards. The systems can cope with both internal and external standard and finally convert the peak size measurement directly to concentration mass or volume units.

4.3.4. Reports

After processing, results can be printed and plotted in a format selected to achieve the desired presentation. By accessing other software, such as a spreadsheet, further treatment such as statistical or cost analysis can be performed. Accessing word processing programs results in the inclusion of chromatograms, tables and graphs for final presentation. Fig. 4.3d shows a typical analysis summary.

Acquisition method	sodaadds	Quantitation method	sodaadds
Units	mg/ml	System number	1
Channel	1	Vial	3
Injection	1	Total injections	1
Run time	4.00 min	Sample rate	5 per sec
Injection volume	20 μl	Sample amount	0.00
Internal standard amt	0.00	Scale factor	0.00
Mode	Analysis	Response factors	Replace
Version			
Description			

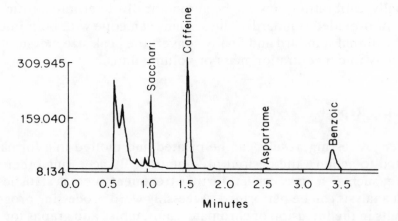

Peak Name	Ret time	Area	Height	Type	Amount	RF
saccharin	1.06	424843	196787	BE	0.152	2.7960e+06
caffeine	1.54	895904	298800	VB	0.085	1.0551e+07
aspartame	2.54	10925	2180	BB	0.057	1.9176e+05
benzoic acid	3.39	329599	53755	BB	0.115	2.8767e+06

Fig. 4.3d.   *Typical analysis summary for the analysis of a diet cola*

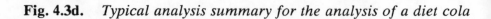

SAQ 4.3a

Fig. 4.3e shows a sample output from a quantitative report.

(*i*) Explain why the normalised concentration should add up to 100%.

(*ii*) Explain how the computer system identifies peak 5 as *d*-guanosine.

(*iii*) What is the purpose of the reference peak at 1.08 minutes?

(*iv*) Explain how the concentration of *d*-guanosine is calculated from the peak area measurement.

```
****** EXTERNAL STANDARD TABLE ******
*************************************************************
*   Sample Name: JT141X
*   Date:  01-01-1980      Method: B:EXTSTD    Operator: 0
*   Interface: 0      Cycle#: 1      Channel#: 1    Vial#:
*************************************************************
Starting delay:  0.00  Stopping retention time:    0.00
Area reject = 10000  One sample per  1.00
Amount injected =  100.00  Dilution factor =    534.84
```

PEAK NUM	RET TIME	PEAK NAME	CONCENTRATION in ug/ml	NORMALIZED CONC%	AREA	HEIGHT	AREA/ HEIGHT	BL	REF PEAK	% DELTA RET TIME	CONC/AREA
1	1.08	Reference Peak	1.06868E+04	13.7380%	4995302	340276	14.7	2	1	0	2.1394E-03
2	3.07	URIDINE	1.12578E+04	14.4721%	7104030	461407	15.4	2	1	-.4159	1.5847E-03
3	4.58	DEOXY-URIDINE	9.59511E+03	12.3347%	3318927	202477	16.4	3	1	-.2351	2.8910E-03
4	5.10	GUANOSINE	9.37642E+03	12.0555%	341859	31152	11.0	4	1	-.5027	2.7428E-02
5	5.52	d-GUANOSINE	9.44689E+03	12.1441%	211956	18628	11.4	4	1	-.5480	4.4570E-02
6	6.70	UNKNOWN	9.18379E+03	11.8059%	6417702	352738	18.2	2	1	-.4562	1.4310E-03
7	7.17	ADENOSINE	9.38505E+03	12.0646%	3500704	204184	17.1	2	1	-7.530	2.6809E-03
8	8.92	CYTOSINE	8.85796E+03	11.3870%	708021	39640	17.9	1	1	-.1209	1.2511E-02

```
TOTAL AMOUNT = 7.77898E+04

PEAKS NOT FOUND IN THIS RUN
NAME        ADJUSTED RET.TIME.    REFERENCE PEAK

GROUP NUMBER      GROUP AMOUNT
      1           2.08529E+04
      2           1.88233E+04
      3           9.18379E+03
      4           1.82430E+04
```

Fig. 4.3e.   *Sample output for a quantitative report*

Learning Objectives

You should now be able to:

- differentiate between the terms qualitative and quantitative analysis as applied to chromatography;

- explain the limitations of chromatography in both qualitative and quantitative analysis;

- list the means by which the identity of a component may be confirmed;

- define the terms relative retention and retention indices, and calculate values from appropriate data;

- explain the response characteristics of chromatographic detectors;

- list available methods for peak size determination and perform the relevant calculations;

- compare methods of quantitative treatment of data;

- calculate sample concentrations given relevant chromatographic data;

- explain the purpose and value of chromatographic data systems.

5.   Classical Column Chromatography

5.1.   INTRODUCTION

At this stage of your study, you will have a detailed knowledge of many of the processes that provide a basis for the separation of mixtures by chromatography. Let us now turn our attention to some of the practical aspects involved in column chromatography.

It was a Russian botanist, M S Tswett, who performed the first scientific investigation using column chromatography in 1906. Tswett reported a very thorough investigation of the separation of chloroplast pigments on over 100 different adsorbent systems.

If you are interested in the historical account, then it would be worthwhile reading a translation of the original paper (J. Chem. Ed. 1967 *44* 235).

Fig. 5.1a shows a diagram of the column chromatographic apparatus used by Tswett. Compare this with a typical modern commercially available system for chromatography, Fig. 5.1b.

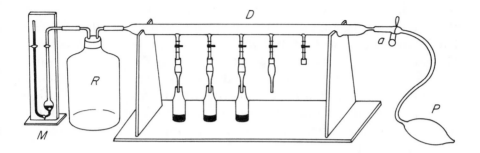

Fig. 5.1a. *A battery of Tswett columns for the simultaneous fractionation of several plant extracts. M, manometer; R, pressure reservoir; D, manifold; a, pinch clamp, P, pump*

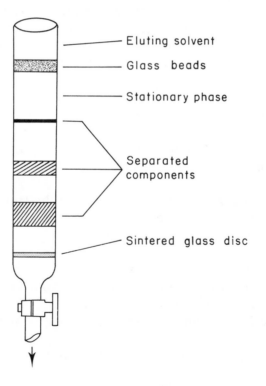

Fig. 5.1b. *Schematic diagram of a typical modern column for chromatography*

In Tswett's experiments, pigments were extracted from leaves using petroleum ether, and the extract applied to the top of a column of a finely powdered, adsorptive material such as powdered chalk. The leaf pigments were initially adsorbed at the top of the column. The solvent was allowed to percolate through the column whilst adding fresh solvent to the top end and, as a result of differential rates of migration and different degrees of adsorption of the various leaf pigments, the components separated into a series of different coloured bands. Notice that pressure is applied to the solvent in Tswett's apparatus.

∏   Can you suggest why pressure was applied to the solvent?

The increased pressure served to increase the flow rate of the solvent through the packed bed of the stationary phase. In fact, a major disadvantage of classical column chromatography is the amount of time needed to perform a separation as compared with the time needed to perform a similar separation using modern liquid chromatography (lc) or thin-layer chromatography (tlc).

Isolation of the individual leaf pigments, once separation had been completed, was performed in two ways:

(*a*) by continuing the flow of solvent until the pigments were eluted from the column and collected separately, *or*

(*b*) by removing the stationary phase from the column and extracting the pigments from each coloured zone of the column using a polar solvent.

The original experiment by Tswett accurately predicts the current techniques of column chromatography, although at first sight there appear to be many differences between his experiment and modern column chromatographic apparatus that are commercially available.

We will deal with stationary phases later in this section.

5.1.2. Commercial Glass Columns

Commercial chromatographic columns are manufactured from glass tubes fitted with a sintered glass frit at the bottom. The glass frit provides a support for the stationary phase. The tube is also fitted with a tap to arrest or control the flow of mobile phase. Teflon taps are superior to glass ones because they are less prone to sticking and leakage. The mobile phase is often held in a reservoir that can be fitted to the top of the column using a ground glass joint.

Various column dimensions have been recommended, the optimum ratio of height to diameter being between 4 and 10.

If you have access to a glassware catalogue from Corning or another manufacturer, you may wish to list the dimensions of the columns commercially available.

The overall size of the column is governed by the amount of stationary phase needed to separate a given amount of sample, since sample capacity increases proportionally with column cross sectional area.

5.1.3. Stages in a Separation

If you have any experience in performing a separation by column chromatography, you will recognise the following stages:

(*a*) Preparation of the stationary phase and packing the column

Stationary phases in this type of chromatography are normally used once then discarded. The packing procedure, that is outlined in the suggested practicals accompanying this unit, is time consuming and needs practice to perform properly. Frequently, as we shall see later, the stationary phase needs pretreatment before it can be used to perform a successful separation.

(*b*) Sample application and separation

Let us assume that the sample is already prepared. Once it has been applied to the top of the column and adsorbed onto the first few millimetres of the stationary phase, mobile phase is fed through the column under gravity.

NB It is important that the top of the bed is not allowed to dry out, otherwise the effectiveness of the stationary phase will be impaired. Provided that the correct selection of mobile and stationary phases have been made, then separation of the sample components will proceed satisfactorily. A typical separation can take several hours.

(*c*) Detection

Finally, the material eluted from the column must be detected and analysed. Small volumes of column effluent are collected and can be analysed chemically or by spectrometric techniques. The results may be plotted graphically in the form of a chromatogram.

∏   Using the above description of a typical separation proce-
       dure, outline two disadvantages that you can identify when
       comparing classical column chromatography to other chro-
       matographic techniques such as, say, tlc.

The factors we have identified as being significant are as follows:

(*a*) the total time required to pack the column, run the chro-
       matogram and detect the components present in the effluent
       makes column chromatography a tedious and time consuming
       exercise, even assuming that automatic detection is used. Can
       you imagine a hard pressed quality control laboratory using
       classical column chromatography for analysing many samples?

(*b*) the size of the columns, the long run times and the fact that the
       stationary phase is rarely re-usable mean that material costs for
       solvents and stationary phases can be high. Coupled with this
       is the additional fact that large volumes of solvent may pose a
       health hazard, a fire risk and may be difficult to dispose of.

Up to this point, classical chromatography does not seem to have a great deal in its favour and this fact is reflected in its limited use in analytical laboratories. So what use is it?

You may remember that we mentioned the amount of stationary phase used in a column depended on the amount of sample to be analysed. The main advantage with column chromatography is that it is possible to scale up the technique to handle large amounts of material. For example, whereas a preparative tlc plate can handle up to 100 mg of material, a 100 cm × 10 cm column can handle approximately 20 g of solid sample. The main use of column chromatography is therefore in preparative and other large scale separations.

SAQ 5.1a

Prior to performing a preparative separation using column chromatography, thin-layer chromatography is often used to check the suitability of a particular mobile and stationary phase combination.

(*i*) Suggest a reason why this procedure is performed in preference to using column chromatography in the first instance.

(*ii*) Why do you think a less polar solvent is often preferred to the chosen tlc solvent for the final column separation?

SAQ 5.1a

SAQ 5.1b   Before we start to discuss the types of stationary and mobile phases used in column chromatography, try to remember the four types of sorption mechanism that provide us with the basis of chromatographic separations. Describe the nature of the stationary phase and how separation of different solutes proceeds. Avoid referring to the beginning of this Unit if possible.

5.1.4. Solvents and their Characteristics

In this section we have restricted our study to solvents used in adsorption chromatography. This is because partition systems are comparatively rare in classical column chromatography, and solvents used in ion-exchange and size exclusion work will be discussed in Section 5.2.

The first part of this section looks qualitatively at the influence of solvents on a separation.

Let us look again at what happens when a solute is adsorbed onto a solid surface. Fig. 5.1c shows a solid surface, typically alumina or silica, which has sites capable of attracting solute molecules. These sites are called 'active' sites.

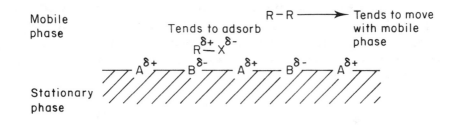

Fig. 5.1c. *Surface of a theoretical adsorbent showing surface sites capable of attracting polar molecules. R-X polar molecule is adsorbed more strongly than non-polar, R-R*

∏ Can you remember the types of force that exist between the solutes and the active sites? If so, name them and give two examples of solutes that would be strongly adsorbed onto an alumina surface.

The intermolecular forces normally responsible for adsorption in chromatographic systems have been classified as follows:-

(*i*) dispersion forces

(*ii*) dipole-dipole forces

(*iii*) hydrogen bonds

(*iv*) weak covalent bonding

Alumina and silica are generally classified as polar adsorbents and will preferentially retain polar molecules. Dipole-dipole interactions will thus be responsible for the retention of alkyl halides on the polar active sites on these adsorbents. Hydrogen bonding can also have a role, as for example in the separation of nitrophenols. 2-nitrophenol is intramolecularly hydrogen bonded, while 4-nitrophenol can hydrogen bond with a polar surface. If they are chromatographed as a mixture on a polar stationary phase, the latter will tend to be held back by interaction with the stationary phase allowing the former to elute first.

Using this mechanism for adsorption, it is easy to see why more polar solutes tend to be held more strongly on polar stationary phases than are less polar solutes. The greater the charge separation on a molecule, the greater the adsorptive forces of attraction. Fig. 5.1d illustrates this interaction between a solute and the stationary phase.

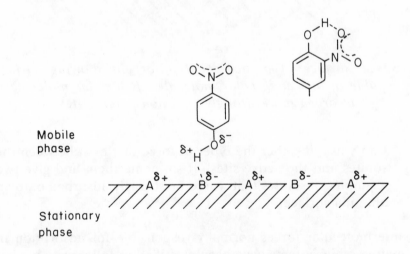

Fig. 5.1d. *Schematic comparison of 2-nitrophenol and 4-nitrophenol on a polar adsorbent surface. 4-nitrophenol hydrogen bonds to the polar surface to a greater extent than 2-nitrophenol. 2-nitrophenol would therefore elute first*

You have to realise that the adsorbent cannot distinguish between the solute and the solvent, ie there is a competition for the adsorption sites between the two, so that a particular combination of solvent and stationary phase must be chosen to give the best separation for a particular mixture.

As a general rule, the problem is to match the polarity of the solvent with that of the sample, whilst using more powerful (active) adsorbents for less-polar solutes and less active adsorbents for more polar substances.

This can be best illustrated by using the following example that is based on one of the suggested practicals for this Unit.

∏   A mixture of three solutes, iodine, methylene blue (a dye) and benzoic acid, are to be separated by column chromatography using an active alumina as stationary phase. Iodine is the least polar and benzoic acid the most.

After placing the mixture on top of the column, tetrachloromethane, a non-polar solvent, is added. Which substance will elute first and why?

The correct answer is iodine. Being less polar than the other solutes it will have less of an affinity for the stationary phase and elute first.

∏   The other two solutes are held strongly at the top of the column. What do you suggest should now be done to continue the separation?

Appropriate action would be to elute with a higher polarity mobile phase. Ethanol, as the second solvent, would successfully elute methylene blue, while the most polar, benzoic acid, would still remain adsorbed quite strongly onto the stationary phase.

∏   How then do we finally obtain the benzoic acid fraction?

Eluting the column with a highly polar solvent such as water, would result in the water molecules displacing benzoic acid molecules from their sites on the stationary phase by virtue of their greater polarity. We therefore have a successful chromatographic separation by selection of appropriate solvents.

This is an example of 'stepwise elution' and illustrates that solvent polarity is an important characteristic to control separations by adsorption chromatography.

SAQ 5.1c

Three aromatic hydrocarbons are to be separated on a highly polar alumina column.

Why is a non-polar solvent such as hexane a more appropriate mobile phase than propanone?

You now know that solvent polarity is an important parameter in selecting the appropriate conditions for a separation.

∏   Can you suggest three other properties of a solvent that could be important?

       Take a few minutes or so to think about this question. Some clues are to be found in the material already included in this section of the text.

Here are our suggestions:

The first point concerns the detection system in use. For example, if solutes are detected by virtue of their ability to absorb ultraviolet radiation at a particular wavelength, methylbenzene as a mobile phase would be inappropriate because its own absorbance would be likely to mask the absorbance of most solutes. Cyclohexane, methanol or acetonitrile would be better propositions.

Secondly, the viscosity of the solvent is an appropriate parameter to consider in classical column chromatography. You know that the time taken to run a chromatogram is relatively long in column chromatography, because of the slow rate at which the solvent percolates through the stationary phase. The flow rate of the solvent at constant pressure is inversely proportional to its viscosity. So, if the viscosity of the solvent is doubled, the flow rate will be halved. Therefore, apart from other considerations, solvents of lowest viscosity should be used.

The third factor concerns solvent toxicity and flammability and is especially important in column chromatography because of the larger volumes involved.

In addition to the above factors, you may have mentioned the cost – many solvents are expensive.

	Polarity ε^o	Viscosity cP	Cost
fluoroalkanes	−0.25		
pentane	0.00	0.24	
hexane	0.00	0.31	
cyclohexane	0.04	1.00	
carbon disulphide	0.15	0.37	
methylbenzene	0.29	0.59	
benzene	0.32	0.65	
ethoxyethane	0.38	0.23	
trichloromethane	0.40	0.57	
tetrahydrofuran	0.45	0.55	
propanone	0.56	0.32	
acetonitrile	0.65	0.37	
propan-2-ol	0.82	2.3	
ethanol	0.88	1.2	
methanol	0.95	0.60	
N,N-dimethylformamide	≫1.00	0.92	
water	≫1.00	1.00	

Fig. 5.1e. *Solvent strengths and viscosities of some solvents currently used in column chromatography*

Fig. 5.1e lists the solvent strengths and viscosities of some solvents currently used in column chromatography. The polarities of the solvents are listed as ϵ^o values, which are derived from solvent strength calculations. The larger the value of ϵ^o the more polar the solvent. ϵ^o has been defined as the adsorption energy per unit area of stationary phase. We have not included the costs of the solvents in Fig. 5.1e because such costs will vary, it would be worthwhile you completing this part of Fig. 5.1e from up-to-date catalogues.

SAQ 5.1d   Assume that you need to separate a four component mixture and that you are provided with
a column filled with an appropriate stationary
phase. After elution with cyclohexane, two components of the mixture separate satisfactorily
and are eluted from the column. However, the
other two components stay firmly adsorbed on
the top of the column. What is your course of
action?

5.1.5.   Fraction Collection

Qualitative and quantitative analysis by classical column chromatography generally involves collecting fractions of effluent from the column and performing a chemical or spectrometric analysis on the collected fractions.

If this is to be done manually, say collecting a total volume of 125 cm^3 in 5 cm^3 fractions, then the procedure is clearly time consuming.

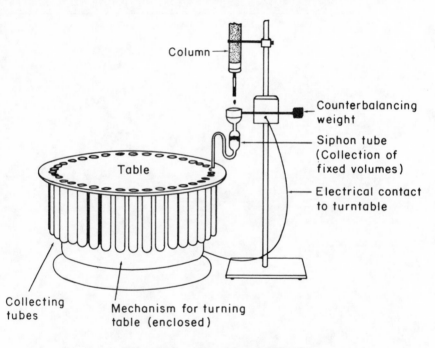

Fig. 5.1f.   *Fraction collector*

An apparatus specifically designed to collect small volumes is shown in Fig. 5.1f. The turntable holds a large number of tubes that are positioned to collect effluent from the column. It is operated electrically by a siphon tube that switches delivery to a new sample tube as soon as a specific mass of effluent has been collected.

However, you will readily realise that the fractions still have to be analysed individually. This is a major limitation on the use of fraction collectors in laboratories.

An alternative to fraction collection is used in modern liquid chromatographs. Effluent from the column is passed through a small volume cell and some property of the effluent is continuously monitored. The changes in the property are measured as a function of effluent volume or time and automatically plotted on a chart recorder. The uv absorbance of the sample or refractive index are typical properties that are measured in continuous monitoring.

5.2. STATIONARY PHASES

This section includes information on the types of stationary phase that are used in column chromatography. The text will concentrate on the stationary phases used in the following types of chromatography:

adsorption chromatography

ion-exchange chromatography

size exclusion chromatography

Stationary phases suitable for use in partition chromatography have been deliberately omitted, because they are rarely used in classical column chromatography, despite their widespread use in gas chromatography.

5.2.1. Adsorbents

Fig. 5.2a lists a selection of stationary phases that have been used as adsorbents in decreasing order of polarity. Two of the list are used far more than others, *viz* alumina and silica.

Highest Polarity	alumina
	magnesium oxide
	carbon
	silica
	magnesium carbonate
	calcium carbonate
	potassium carbonate
	sodium carbonate
	starch
Lowest Polarity	cellulose

Fig. 5.2a.   *List of Adsorbents used in Classical Column Chromatography*

∏   Notice that no numerical values related to polarity have been assigned to the individual materials. Can you suggest why it is impossible to assign a numerical value?

This question is in fact a very difficult one, despite attempts in the literature to answer it. It is very difficult to specify what acts as an active site in practically any adsorbent. Many texts suggest the effect is purely a physical interaction between surface groups on the surface of the adsorbent and polar molecules in solution. However, it is likely that the nature of the surface in terms of its defect properties has a role to play. When we study specific adsorbents, we will also find that the presence of other materials on the surface (called adsorbates) have a marked influence on surface polarity.

As a result, you must accept that the order in which the adsorbents are listed may alter slightly under different operating conditions.

Can we classify adsorbents in any other way?

The answer is yes. In addition to polarity, we can describe some adsorbent phases as *acidic* and others as *basic*.

Let us look at a specific example of each type.

Silica has the ability preferentially to retain amines and other basic compounds compared with non-basic compounds of a similar polarity. For obvious reasons it is therefore described as *acidic*.

Conversely, alumina and magnesia preferentially retain acidic compounds and are described as *basic* adsorbents.

∏   In addition to the above classification, there are other requirements with which an adsorbent has to conform. It is worth spending a few minutes considering what these requirements may be. You may not get all the points, but getting even one or two will be a help.

We managed to think of the following. They are not listed in order of importance:

(*i*) Adequate particle strength

This has a number of alternative names, such as 'resistance to attrition' or even 'friability'. When columns are packed, particles of adsorbent can be subjected to damage. This can affect the effectiveness of the separation. More importantly, the column itself can become blocked.

(*ii*) Reproducibility of surface characteristics

Firstly, the chemical nature of the surface should be well defined. Although perhaps not so important for a 'one-off' separation, for repeated work requiring reproducible separations it becomes important. Related to this fact is the physical nature of the surface, and particularly the surface area. High surface areas are necessary for efficient separations, and the ability of a given mass of adsorbent to separate an optimum mass of solute. This latter property is termed the *linear capacity* of an adsorbent.

(*iii*) Chemical Stability

There are two ways in which chemical stability is important. Firstly, the stationary phase should be resistant to chemical attack. Secondly, despite the fact that solutes are adsorbed, the adsorbent should not catalyse chemical changes in the adsorbate.

SAQ 5.2a	Explain the terms 'linear capacity' and 'theoretical plate'.

5.2.2. Particle Size of Adsorbents

It is worth us looking at the implications of variations in the particle size of stationary phases in more depth. Many parameters related to separation are closely dependent on particle size.

Fig. 5.2b shows the relation between the average particle size of an adsorbent and the number of theoretical plates, which in turn is a measure of the efficiency of a column (see Part 3).

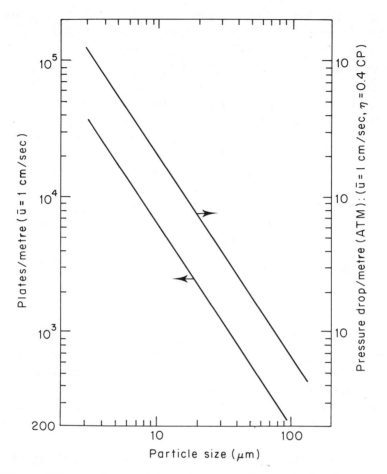

Fig. 5.2b. *Effect of particle size on theoretical plates and column pressure drop*

You can infer from this diagram that the smaller the particle size, the more efficient the separation. There is, however, a practical limitation to the minimum particle size of the stationary phase that is used in classical chromatography.

Π   Think for a few minutes about this problem and state what you think this limitation is.

As the particle size decreases, column efficiency increases, but there will also be an increase in the resistance to mobile phase flow through the bed. High performance liquid chromatography is far more efficient than classical column chromatography because it is possible to use very small particles, but high pressures are necessary to force the mobile phase through the densely packed hplc columns.

Fig. 5.2c shows typical sizes of adsorbents used in both classical column and high performance liquid chromatography.

Technique	Particle Size
Classical column	0.06–0.200 mm
hplc	3–20 μm

Fig. 5.2c. *Particle sizes for classical and high performance liquid chromatography*

The particles sizes discussed above are normally quoted as an average. Unfortunately, stationary phases cannot be produced with all the particles having identical diameters. There is always a particle size distribution which should, ideally, be spread over as narrow a range as possible for optimum packing into columns and efficient chromatography.

These points can be summarised in the following statements:

larger particles limit column efficiency

smaller particles limit column permeability

The following is a passage written in 1941 by two chromatographers who worked on the theoretical basis of partition chromatography:

'... the smallest height equivalent of a theoretical plate (hetp) should be obtainable by using small particles and high pressure differences across the length of the column."

It was almost 30 years before their suggestion was adopted by manu-
facturers of the first commercial systems for liquid chromatography.

5.2.3. Additional Adsorbent Characteristics

Perhaps the most difficult property of an adsorbent to characterise
is its pore size distribution.

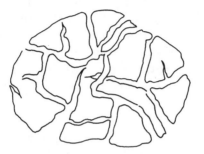

Fig. 5.2d. *Porous nature of adsorbent particles such as silica*

Fig. 5.2d. shows a schematic diagram of pores that may be present in
particles of, for example, silica. The presence of the pores has a pro-
found influence on its ability to separate components of mixtures.
The silica has a range of pore sizes and shapes, which depends on
the way in which it was made. Perhaps the only general comment
that can be made is that for most purposes pore sizes must be greater
than 5 nm to prevent irreversible adsorption.

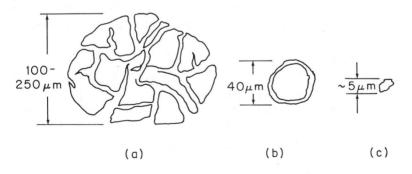

(a)   (b)   (c)

Fig. 5.2e. *Comparison of adsorbent stationary phase particles*

(*a*) classical column chromatography

(*b*) pellicular stationary phase, porous shell between 1 and 3 μm thick

(*c*) microporous particles of diameter $\sim$5 μm for hplc

A particle of a typical column packing for classical liquid chromatography is illustrated schematically in Fig. 5.2e. The average diameter of the particles is between 0.1 and 0.25 mm. Very little pressure is required to allow solvent flow between these large particles, but separation times are long because of the slow flow rates. The presence of deep narrow pores in these particles can lead to slow desorption or irreversible adsorption of solutes within the bulk of the particle.

Early developments in particle technology led to the introduction of new types of stationary phase. In order to overcome the limited mechanical strength of porous particles, and speed up desorption from deep pores, 'pellicular' particles were developed. They have a smaller particle diameter giving increased column efficiencies. The particle consists of a solid non-porous core, usually a glass bead with a diameter of approximately 40 μm. This is coated with a porous shell 1 to 3 μm thick which acts as the adsorbent. These pellicular particles are easily packed into a column, but sample capacity is low due to a low value of V_S (stationary phase volume).

For high performance liquid chromatography, microparticles in the range 3 to 20 μm are used which give very high efficiency columns. They have fast separation times and moderate capacities, but are comparatively expensive.

We have now studied some of the important aspects of adsorbents that influence chromatographic separations. The inter-relation between surface area, surface properties, particle size, particle size range, pore size and pore size range means that their individual and combined effects on a separation are complex and often difficult to predict.

5.2.4. Specific Adsorbents

Silica and alumina are the most popular adsorbents used in classical column chromatography. These adsorbents are normally used in the form of coarse, porous particles. The higher surface area associated with porous particles means that the efficiency of the column is maximised with an acceptable solvent flow rate at low pressure.

Heat treatment of silica results in changes in the adsorptive properties of the surface. For example, heating the surface to above 200 °C leads to a loss of ability to separate polar compounds.

If the surface is allowed to adsorb various amounts of water, then the surface activity can be altered over a considerable range. The polarity is a maximum when silica that has been exposed to water is then heated to approximately 170 °C.

∏ Suggest how the surface is modified both by heat treatment and wetting. Try to avoid using other texts for this exercise.

Did you find this a difficult exercise to approach? The answer is not difficult to understand, but perhaps more difficult to explain. You will have the chance to perform a similar exercise later for alumina. The answer illustrates how adsorptive properties are dependent on the chemical nature of the surface.

The adsorptive properties of silica depend on the presence of surface silanol (Si—OH) groups. Solute molecules are adsorbed as a result of the formation of hydrogen bonds, the silanol groups acting as hydrogen donors.

At about 170 °C, the polarity of the surface is a maximum because physically adsorbed water, generally present on the surface, is removed and the maximum number of silanol groups are present. Exposure of this surface to a low vapour pressure of water increases the amount of physically adsorbed water on the surface. This coating decreases the interaction between the solute molecules and the silanol groups.

Control of the amount of surface water is thus a means of varying the polarity of the surface.

Heating silicas above 200 °C leads to the conversion of silanol groups to siloxane groups (Si—O—Si). This leads to a decrease in the ability of the surface to retain polar materials. A schematic diagram of the siloxane surface is shown in Figure 5.2f.

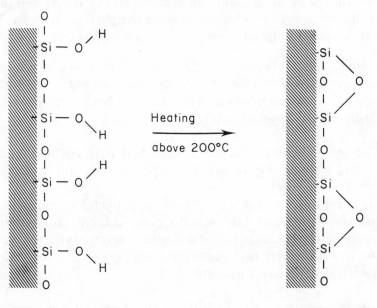

Active surface   Inactive surface

Fig. 5.2f.   *Conversion of a hydrated silica surface to siloxanes by heating*

In addition to interaction between surface groups and solutes via hydrogen-bonding, the silica surface is mildly acidic. Amines and other bases tend therefore to be preferentially retained.

When you read about the characteristics and requirements of stationary phases, an important consideration was that the surface should not catalyse chemical changes in solute molecules. Silica has little tendency to do this.

Alumina used for chromatography exists in a specific crystalline modification called gamma-alumina (γ-Al_2O_3). This alumina forms the basis of all 'activated aluminas' which are open-structured materials widely used as catalysts and chromatographic stationary phases. The alumina surface shows similar adsorbent properties to silica, but the fact that acids are preferentially retained on alumina surfaces, suggests there are some chemical differences in the surface groups of alumina and silica. The precise nature of the chemical groups that act as adsorbent sites has not been identified, although its ability to retain acids is probably due to interaction with the basic oxide sites on the surface.

Heating alumina to above 1000 °C results in a total loss of adsorptive properties. At this temperature, the gamma form is converted to the chemically inert, close packed alpha form. Nevertheless, heat treatment at between 200–400 °C followed by water deactivation, produces a good general purpose adsorbent.

5.2.5. Standarization of Adsorbents

If consistent separations need to be performed, it is important to prepare stationary phases as reproducibly as possible.

You should now recognise the difficulty in defining precisely the nature of a surface. Nevertheless, methods have been developed that allow polarities of surfaces to be compared on an arbitrary scale. A measure of polarity of a silica or alumina surface is its activity. The *Brockmann activity scale* defines five grades ranging from fully active (I) to least active (V). The particular activity required for a separation may be selected by adding a measured amount of water.

Fig. 5.2g summarises the Brockmann activity scale for commercial forms of both silica and alumina.

	\multicolumn{5}{c}{Percentage of added water}				
	0%	20%	40%	60%	80%
Silica Gel 40	I	II	III	IV/V	V
Silica Gel 60	I/II	II	III	III/IV	V
Silica Gel 100	II	II/III	III	III/IV	IV
Alumina 60	I	II/III	III/IV	IV/V	V
Alumina 90	I/II	III	III/IV	IV	V
Alumina 150	II	III/IV	IV	IV	IV/V

Fig. 5.2g. *Brockmann activity scale for commercial forms of alumina and silica*

5.2.6. Ion-exchangers

This section deals with the specific application of column chromatography to the separation of ions, although the technique of ion-exchange has wider applications than simply separation.

You have already been introduced to the basic mechanism of ion-exchange, (Section 1.5.4), but will not yet have identified the specific nature of the stationary phases used.

Historically, the first ion-exchange materials used in chromatography were natural materials such as clays and zeolites. The introduction of polymer-based ion-exchange resins meant that materials could be manufactured with more reproducible characteristics in terms of size and ion-exchange capacity.

Flow rates for ion-exchange columns must be comparable with flow rates for adsorbent columns.

∏   Can you remember the particle size range suitable for classical column chromatography using adsorbents?

To ensure sufficient flow and reasonable retention times, particle diameters between 0.06 and 0.20mm are suitable.

Many different ion-exchange materials have been used in column chromatography. One of the early types still in use is based on polystyrene. However, these materials are hydrophobic and present problems when used to separate very large ionic molecules. Cellulose-based ion-exchange materials provide a hydrophilic alternative.

Hydrophobic Ion-exchangers

The structure of an ion-exchange resin is polymeric and there are actually two basic types in common use, one of which is based on polystyrene and the other on polymethyl-methacrylate (pmma). Both types contain a percentage (up to 20%) of a cross-linking agent called divinylbenzene (dvb).

The percentage of dvb added to the polymerisation mixture is defined as the cross-linking percentage. Cross-linking gives the polymer a higher degree of rigidity and increases its resistance to mechanical damage. You should try to remember this definition and explain why it is important for stationary phases to have mechanical strength.

However, a high degree of cross-linking reduces the porosity of a resin which results in a lower rate of exchange of ions and limits the total capacity of the resin in terms of its sample loading. Such a resin does not, however, swell appreciably in the presence of solvents, thus making it easy to maintain a properly packed column.

The polymerisation process produces porous, spherical particles of the resin, whose particle size can be controlled within a narrow range.

Schematic diagrams of both types of resin are shown in Fig. 5.2h.

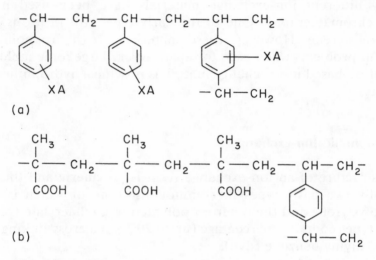

Fig. 5.2h. *Formulae of typical ion-exchange resins*

(a) polystyrene – dvb. X represents the fixed
    ion, A the counter ion.

(b) polymethylmethacrylate (pmma)

Compare the structures of the two resin types. You will observe that
the pmma resins are carboxylic acids. These are weakly acidic, but
the hydrogens can be replaced by other cations that are in contact
with it.

There are no ionic groups in polystyrene, so they must be introduced
chemically after polymerisation.

∏   What types of group could be introduced into the resin?

You may be able to answer this question from prior knowledge.
If not, obtain a chemical suppliers catalogue or other convenient source of reference. Remember that both cation and anion-
exchangers are available, so that at least two types of substituent
group must be introduced into the resin matrix.

You should have found reference to four types, as follows:

Cation-exchangers: strongly acidic   sulphonic acid

$$-SO_3^-$$

weakly acidic   carboxylate

$$-COO^-$$

Anion-exchangers: strongly basic   quaternary ammonium

$$-N^+R_3$$

weakly basic   tertiary ammonium

$$-N^+R_2H$$

You may have found reference to some other types such as chelating resins. These materials will be discussed later in the text.

Let us look at the process of ion-exchange that occurs with one of these resin types. The process is similar for all resins so that once you have understood one, you can apply the same reasoning to the others. You will have to remember some definitions from this section.

At the start of this Unit you were introduced to the basic process of ion-exchange. Here is one definition of the process:

It is the reversible interchange of ions between an insoluble solid phase (the ion-exchanger) and a solution phase.

The definition is easy to remember, but don't forget the key word, *reversible*. Ion-exchange is an equilibrium process; you will recognise the importance of this in the following section.

Fig. 5.2i shows a schematic diagram of an ion-exchange resin in contact with a solution containing ions. We have chosen to illustrate this example with a strongly acidic cation-exchanger in the hydrogen form.

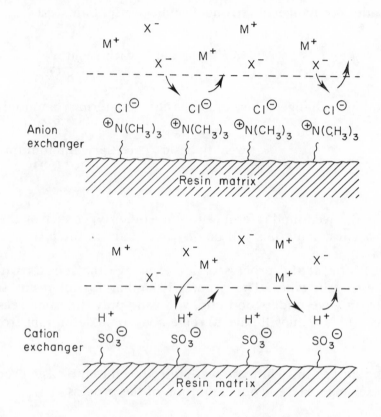

Fig. 5.2i. *Schematic diagram of ion-exchange resins in equilibrium with a solution of ions*

The commonest retention mechanism is simple ion-exchange as represented by the following equation:

$$R-SO_3^-/H^+ \quad + \quad M^+ \quad = \quad R-SO_3^-/M^+ \quad + \quad H^+$$

solid   solution   solid   solution

A strongly acidic ion-exchange resin will have a sulphonic acid group that is completely ionised over the pH range 1 to 14. In general, the cations in solution will have a greater affinity for the resin than the hydrogen ions and will displace them.

At some stage in any exchange process, the resin must become 'exhausted', ie the point when all possible exchange sites have been replaced by ions from the solution. This property is termed the *capacity* of an ion-exchange resin.

It is defined as follows:

The capacity of an ion-exchange is the amount of exchangeable univalent ion per gram of dry resin.

The capacity will depend on the type of resin used, but approximate figures for typical commercial ion-exchange resins are given below:

$$\text{weakly acidic} \qquad 10^{-2} \text{ mol g}^{-1}$$
$$\text{strongly acidic} \quad 5 \times 10^{-3} \text{ mol g}^{-1}$$
$$\text{strongly basic} \quad 4 \times 10^{-3} \text{ mol g}^{-1}$$

In practice, if the hydrogen ions of 100 g of a sulphonic acid resin in the hydrogen form were replaced completely, the free acidity obtained in the solution would be $100 \times 5 \times 10^{-3}$ mol.

This information may be used to calculate the capacity of an ion-exchange resin. The following SAQ demonstrates this calculation.

SAQ 5.2b   0.5462 g of dry ion-exchange resin in the sulphonic acid form were transferred to a 100 cm^3 beaker with 100cm^3 of deionised water. An excess, approximately 2 g, of Analar sodium chloride, was added, allowed to equilibrate with the resin and then the total volume titrated against standard, approximately 0.1 mol dm^{-3}, sodium hydroxide solution using screened methyl orange as indicator. $\longrightarrow$

SAQ 5.2b
(cont.)

As the base was added, the indicator changed colour at what appeared to be the end-point of the reaction. However, after a short time, the original colour of the indicator was regenerated. Additional volumes of base were necessary to ensure that the end-point reached was stable.

Total volume of base added $= 24.50$ cm^3

Molarity of base $= 0.1051$ mol dm^{-3}

You may use these results to answer the following:

(*i*) Calculate the capacity of the cation-exchange resin.

(*ii*) Suggest why several end-points were observed before the end-point colour remained stable.

SAQ 5.2b

Hydrophilic Ion-exchangers

Ion-exchangers based on cellulose and dextrans are hydrophilic and have large porous structures that make them particularly suitable for separations of large ions such as proteins and nucleic acids. The fibres can be chemically treated by an analogous process to that for hydrophobic resins to introduce strong and weak acidic and basic groups into the matrix. Dextran ion-exchangers have the commercial name 'Sephadex'.

Typical commercial cellulose-based ion-exchangers are listed in Fig. 5.2j. Their ion-exchange capacities $(1 \times 10^{-3} \text{ mol g}^{-1})$ are lower than the equivalent ones for resin ion-exchangers.

Anion Exchangers		Ionisable Group
AE cellulose	aminoethyl	$-O-CH_2-CH_2-NH_2$
DEAE cellulose	diethylaminoethyl	$-O-CH_2-CH_2-N(C_2H_5)_2$
TEAE cellulose	triethylaminoethyl	$-O-CH_2-CH_2-N(C_2H_5)_3$
PAB cellulose	*p*-aminobenzyl	$-O-CH_2-C_6H_4-NH_2$

Cation Exchangers		Ionisable Group
CM cellulose	carboxymethyl	$-O-CH_2-COOH$
P cellulose	phosphate	$-O-\overset{\displaystyle O}{\overset{\|}{P}}-OH$ with OH
SE cellulose	sulphoethyl	$-O-CH_2-CH_2-\overset{\displaystyle O}{\underset{\displaystyle O}{\overset{\|}{\underset{\|}{S}}}}-OH$

Fig. 5.2j. *Examples of some commercial cellulose-based ion-exchangers*

5.2.7. Retention Parameters

Let us investigate how we may evaluate retention parameters for ion-exchange systems.

A distribution coefficient, K_m, indicative of the affinity of a specific ion for the resin can be defined as follows:

$$K_m = \frac{\text{amount of ion in resin}}{\text{amount of ion in solution}} \qquad (5.1)$$

A second distribution ratio may also be defined:

$$K_g = \frac{\text{amount of ion in resin per unit mass of resin}}{\text{amount of ion in solution per unit volume of solution}} \qquad (5.2)$$

(This is similar to the equation given for partition in Section 1.5.2)

Therefore
$$K_g = \frac{K_m V_m}{m_s} \qquad (5.3)$$

where V_m is the mobile phase or void volume and m_s is the mass of the stationary phase. If K_g is multiplied by the density of the resin d, then an ion-exchange distribution coefficient K is defined and Eq. 5.3 becomes:

$$K = \frac{K_m V_m}{V_s} \qquad (5.4)$$

Eq. 5.4 leads to another equation that relates the retention volume V_R to the distribution ratio, stationary phase volume V_s and the mobile phase or void volume V_m.

∏   Either derive this relation for yourself, or refer back to Section 2.1.1.

The equation that relates the retention volume V_R to the distribution coeffient, K, stationary phase volume V_s and the void volume V_m is given by:

$$V_R = V_m + K V_s \qquad (5.5)$$

(see also Eq. 2.14)

We can use Eq. 5.5 to calculate retention volumes for ion-exchange separations in the same way as we use it to calculate retention volumes in partition chromatography.

Now try this problem.

∏   In an ion-exchange chromatographic separation of two ions, their distribution coefficient were 1.05 and 3.85 respectively. Given that the stationary phase volume (V_s) is 10 cm^3 and the void volume (V_m) 7 cm^3, calculate the retention volumes of each of the ions.

You should use Eq. 5.5,

$$V_R = V_m + KV_s$$

For the first ion K is 1.05, V_s is 10 and V_m 7

Substitution in Eq. 5.5 gives

$$V_R = 1.05(10) + 7$$
$$= 17.5 \text{ cm}^3$$

Similarly the retention volume for the second ion is given by

$$V_R = 3.85(10) + 7$$
$$= 45.5 \text{ cm}^3$$

Influence of the Mobile Phase

In the problem you have just completed, the affinity of each ion for the stationary phase was measured in terms of a distribution coefficient. Let us now turn our attention to ways in which varying the composition of the mobile phase can influence the separation of solutes.

Earlier in this section, we mentioned that chelating resins were available that could perform as stationary phases. The most common stationary phase in this category is a resin that has iminodiacetic groups attached:

$$\begin{array}{c} CH_2COOH \\ | \\ -CH_2-N \\ | \\ CH_2COOH \end{array}$$

As you should be aware, complexing agents have an affinity for metal ions because of their ability to donate electrons to a cation. You may have come across common chelating agents such as 1,2-diaminoethane and EDTA.

Chelating resins, as ion-exchangers, have an affinity for metal ions, particularly those of heavy metals, and are important in removing these metal ions from solution.

Given the basic types of ion-exchanger – anion, cation and chelating, let us see how we can exert some control over the separations of ionic species by altering the composition of the mobile phase.

In general, we may outline three methods:

(*i*)   use of complexing (usually chelating) agents

(*ii*) adjustment of ionic strength

(*iii*) adjustment of pH

Combinations of these methods, such as the simultaneous alteration of pH and use of complexing agents have also been used.

Use of Complexing Agents

The classic use of complexing agents was in the separation of chemically similar elements of the actinide and lanthanide groups. A column is eluted with a solution of a complexing agent, such as citrate. The groups on the stationary phase and the complexing agent in the mobile phase compete for the metal ions present. The greater the degree of affinity of the metal ion for the complexing agent in the mobile phase the faster it will be eluted. Such a system allows a much higher degree of control over the separation parameters, than would be possible in the absence of complexing agents.

Adjustment of Ionic Strength

Suppose that some solute ions are dissolved in a buffer and that this solution is added to an ion-exchange column.

We may assume that the surface of an ion-exchanger presents a random distribution of charge to the ions being retained. If we assume a purely electrostatic interaction, the effectiveness of the surface in retaining ions will be determined by the surface structure and the competition between the buffer ions and the solute ions for the sites.

Let us suppose that the buffer solution carrying the solute ions has a low salt concentration, ie its ionic strength is low. In this situation, the ions from the buffer do not present significant competition for the sites and the solute ions will be attracted to the stationary phase.

If the salt concentration (ionic strength) of the buffer is increased, the increased competition offered by the buffer ions may be sufficient to displace the solute ions from the stationary phase resulting in the ions tending to leave the stationary phase and enter the mobile phase. At this point, the solute ions will start to move down the column.

Different ions will have different affinities for the stationary phase; the effect of gradually changing the ionic strength will often result in ions being displaced sequentially from the stationary phase according to their degree of affinity.

Although there is an analogy here with adsorption chromatography, where the polarity of the solvent is changed, the mechanism in ion-exchange differs in that once an ion is displaced it tends not to equilibrate between the mobile and stationary phase, but remains in the mobile phase.

The technique of gradually changing the ionic strength of the mobile phase is called *gradient elution*. The limit of the gradient (the highest ionic strength) should be sufficient to elute the most tightly bound solute.

The alternative to gradient elution is *stepwise elution* in which a succession of different solutions, each stronger than the previous, are used to elute the solute ions. Stepwise elution is not so versatile as gradient elution since a number of different ions, each having approximately equivalent affinities for a mobile phase may co-elute as a single peak. There is more control over a separation by adopting the gradient method. For this reason and the simplicity of the necessary apparatus, stepwise elution is limited to the analysis of samples containing few component ions.

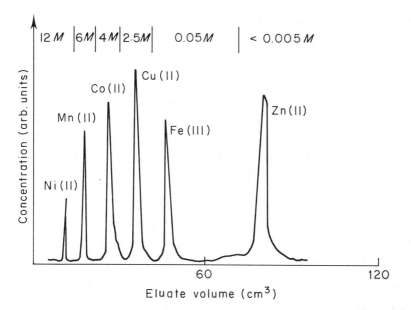

Fig. 5.2k. *Separation of transition elements (Mn to Zn) by stepwise elution with increasing concentrations of HCl. Column: Dowex cation-exchange resin*

Fig. 5.2k shows a chromatogram of an cation-exchange separation of a series of transition elements by stepwise elution. Notice the varying ionic strength of the eluting solvent (hydrochloric acid).

Adjustment of pH

Consider the ion-exchange separation of weak acids. An equilibrium is established as follows, where HA is a weak acid:

$$HA \rightleftharpoons H^+ + A^-$$

As pH falls, you should recognise that the equilibrium will shift to the left – this will decrease the number of anions in solution.

Thus, the retention of anions on an anion exchange column would be suppressed. An increase in pH would give rise to the converse situation. The position of the equilibrium will depend on the nature of the weak acid. It is evident, therefore, that adjustment of pH will have an influence on separation parameters for the case of weak acids. A similar situation arises for weak bases. You should try to explain what influence change of pH will have on weak bases before you continue.

Proteins and nucleic acids are examples of species that can be separated by ion-exchange. Both classes of compound can exist as ions, having acidic and basic groups on the same molecule. You may be aware that amino acids possess similar properties; they can be protonated in acid solution thereby becoming cationic species or they can donate their protons to stronger bases and become anionic.

In an ion-exchange separation of proteins, for example pH can be a potent factor in the affinity of a particular protein for the resin and hence its retention time. Increase in pH is expected to increase the net negative charge on a protein that contains an ionisable carboxylic acid group. The affinity of such a protein for a cation-exchanger will therefore decrease, while the affinity of the same protein for an anion-exchanger will be enhanced.

The result of changing the pH of the mobile phase for proteins can thus be equivalent to the change in ionic strength of the mobile phase for simple ions.

The general practice adopted for the separation of proteins on cellulose-based anion-exchangers is to apply the sample at as high a pH value as possible, and then to elute with a gradient of decreasing pH.

Conversely, equivalent samples are applied to cation-exchangers at low pH, the components then being eluted with a gradient of increasing pH. It is often advantageous to increase the ionic strength simultaneously. Think for a few minutes about the possible advantage in increasing the ionic strength of the mobile phase as the separation proceeds.

It is increased for the same reason as in the case of simple ions. You may recall that the ionic strength gradient has the effect of reducing the affinity of an ion for the stationary phase. This reduces the magnitude of the pH required to elute a particular ion.

Now that we have completed the section on ion-exchange, try the following SAQ.

SAQ 5.2c   The first three fractions, called A, B and C respectively, separated from a sample of egg white on a carboxymethyl cellulose ion-exchange column had retention volumes as follows:

$$A \quad 22 \text{ cm}^3$$
$$B \quad 48 \text{ cm}^3$$
$$C \quad 76 \text{ cm}^3$$

Given that the volume of the ion-exchanger in the column is 10 cm^3 and that the void volume is 5 cm^3, calculate the distribution coefficient of each of the components A, B and C.

SAQ 5.2c

SAQ 5.2d
After fractions A, B and C, described in SAQ 5.2c were eluted, the pH of the buffer was increased from its original value of 4.8 to elute other components, at the same time increasing the ionic strength. Discuss the effect of this increasing pH and ionic strength on ions held by the stationary phase.

SAQ 5.2d

5.2.8. Size Exclusion Materials

This Section deals with a chromatographic technique that has been variously described as *gel permeation, gel filtration* and *molecular sieve chromatography*. Size exclusion chromatography (sec) is perhaps a better and more general description.

The basis of the technique lies, not in the chemical interaction of molecules or ions with a stationary phase, but on the size and shape of molecules.

The stationary phase is designed so that molecules can enter pores present in their structure. Smaller molecules diffuse into the structure faster and further than larger ones and can enter smaller diameter pores. As a result, they are preferentially retarded by the stationary phase. Molecules that are larger than the biggest pore diameter are excluded from entering the pores, but are free to travel in the channels between them.

The technique of size exclusion chromatography is used for characterization of monomeric and polymeric species with a relative molecular mass between about 400 and 3 000 000.

Dextran gels are common stationary phases for sec; they are manufactured under the trade name of Sephadex. They are made from branched polysaccharides, which are composed exclusively of glucosyl residues. These chains are cross-linked to produce a porous structure with a pore size distribution that is defined by the extent of the cross-linking. The gels are hydrophilic and have been widely used in biochemical analysis, since the separation of biological macromolecules has to be performed in aqueous solution. However, these gels tend to be soft and also swell or shrink in the presence of certain mobile phases.

Π   What do you consider to be the effect of stationary phase swelling during a column separation?

You should recognise that such swelling can slow the flow rate of the mobile phase, making fast efficient separations impossible, but perhaps more importantly making reproducible separations virtually impossible.

In order to satisfy the demand for rigid hydrophilic sec media, modified silica gels are available. These silicas have a porosity controlled by their method of manufacture. They are modified chemically by bonding additional groups onto the surface (see Section 1.5.3). A typical example is shown below:

$$-Si-(CH_2)_3-O-CH_2-\overset{\displaystyle OH}{\overset{\displaystyle |}{C}}H-CH_2OH$$

But what about separations in hydrophobic media?

The solution is to use a stationary phase that we are already familiar with. In the previous section on ion-exchange, you remember we mentioned the hydrophobic nature of styrene-divinylbenzene polymers which makes them suitable for use with organic solvents. These can be manufactured with different degrees of porosity and no ion-exchange sites thus providing effective sec stationary phases. Materials are available that are able to separate mixtures with a relative molecular mass range of approximately 50 to 10^7.

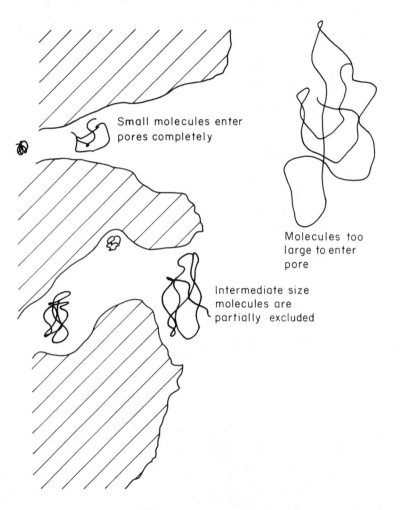

Fig. 5.21. *Schematic illustration of the physical basis of separation using a size exclusion stationary phase. The porous structure preferentially retains smaller molecules*

Fig. 5.21 is a schematic diagram that illustrates the principle of separation on the basis of size. Sec is the simplest of all the sorption mechanisms to understand.

Π   A gel permeation separation of a set of polystyrene standards
     of known relative molecular mass gave the following results
     of retention volume as a function of relative molecular mass.

Relative Molecular Mass	Retention Volume/cm^3
1 000	100
7 000	100
15 800	95
25 000	90
100 000	76
316 000	64
2×10^6	45
10^7	33
$>10^8$	30.

(*i*) Plot the log of the relative molecular mass against retention
       volume.

(*ii*) Establish the significance of the regions of the graph that have
       values of retention volumes $100 > V_R > 30$

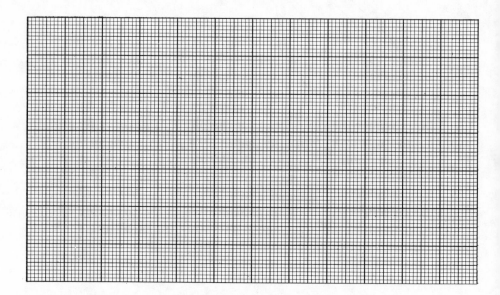

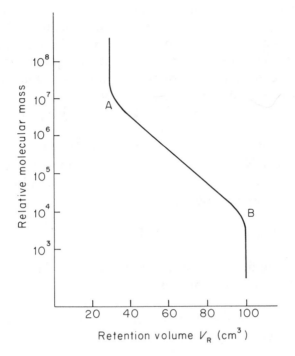

Fig. 5.2m. *Plot of relative molecular mass of a series of polystyrene standards against retention volume*

(*i*) Fig. 5.2m shows the plot of the log of the relative molecular mass against retention volume.

(*ii*) The lower value of the retention volume, point A on the graph, corresponds to a relative molecular mass, and hence size of molecule, that is not retained by the stationary phase. This point corresponds to the *total exclusion limit* and shows that molecules with a relative molecular mass $> 10^7$ are too large to enter the pores. They are not retained, and elute quickly from the column.

Point B, is called the *total permeation limit* for the stationary phase, and shows that molecules of relative molecular mass $< 10\,000$ are small enough to permeate the stationary phase completely.

The region between these two values is the region of *selective permeation* where separation of solutes as a function of their molecular size and shape takes place. Separation is achieved by diffusion of molecules (which always remain in the mobile phase) into the pores of the stationary phase. The range of sample relative molecular masses that can be satisfactorily separated will depend on pore size and on pore size distribution; different grades of stationary phase have to be used for different ranges of relative molecular mass.

Let us see how we can use a technique for characterising sec stationary phases. You have come across the following equation before. It relates the retention volume V_R to the void volume V_m and the stationary phase volume V_s.

$$V_R = V_m + KV_s \qquad (5.6)$$

where K is the distribution coefficient for a solute between the mobile and the stationary phases.

If a molecule is too large to enter the pores, its distribution coefficient, K, is zero.

Eq. 5.6 then becomes

$$V_R = V_m \qquad (5.7)$$

ie the retention volume is equivalent to the volume of mobile phase held within the column packing, the void volume. For species that are small enough to enter all the pores, the distribution coefficient has a value of unity.

Eq. 5.6 then becomes

$$V_R = V_m + V_s \qquad (5.8)$$

and $\qquad V_s = V_R - V_m \qquad (5.9)$

Use these equations to calculate the characteristics of the stationary phase in the following SAQ:

SAQ 5.2e

> A molecular species of relative molecular mass 120 000, which was eluted through a stationary phase of total exclusion limit 80 000, had a retention volume of 25 cm^3.
>
> A material of low relative molecular mass had a retention volume of 225 cm^3, and a sample of unknown relative molecular mass had a retention volume of 125 cm^3.
>
> Calculate the void volume, stationary phase volume and the distribution coefficient for the sample component of unknown relative molecular mass.

A practical illustration of sec is shown in Fig. 5.2n. The sample is a series of polystyrene standards of different relative molecular mass separated on a styrene divinylbenzene stationary phase. Obviously, the stationary phase in this particular separation is effective for the range of relative molecular masses of the polystyrene standards. How effective would the same stationary phase be for samples of different molecular mass range? Clearly, the average pore size will be an important characteristic.

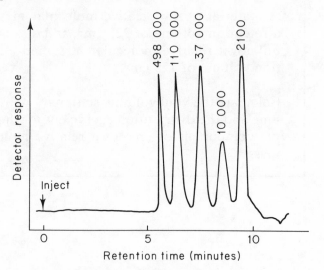

Fig. 5.2n. *High speed exclusion chromatogram of polystyrene molecular mass standards*

∏   The calibration curve shown in Fig. 5.2o shows the effect of varying the pore size of a specific type of stationary phase on the retention volume of standards.

Assume that the following polystyrene samples are to be separated:

polymer 1, $M_r = $ 12 600

polymer 2, $M_r = $ 45 000

polymer 3, $M_r = $ 69 000

polymer 4, $M_r = $ 160 000

Use Fig. 5.2o to choose an appropriate stationary phase for the separation.

Relative molecular mass of polystyrene

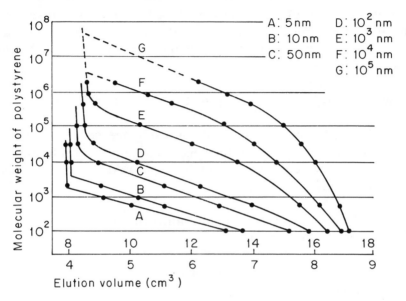

Upper scale: MICROGEL 50 cm columns
Lower scale: MICROGEL 25 cm columns

Fig. 5.2o. *Calibration curves for stationary phases of varying pore size*

Our choice of stationary phase would be E, because it would separate the particular range of molecular mass we are interested in. If you take the log of the relative molecular mass and use curve E to read off the appropriate value of the retention volume, you should get values of approximately 9.8, 11.6, 12.6 and 13.8, using a 50cm column, ie values that lie within the region of selective permeation.

5.3. APPLICATION OF ION-EXCHANGE AND SIZE EXCLUSION CHROMATOGRAPHY

You now have more than a working knowledge of chromatographic separations. In order to complete this Unit, we will turn our attention to the application of chromatography to real situations, in particular those where the mechanism of separation is size exclusion or ion-exchange.

We will start by examining applications of ion exchange.

∏   You remember that true ion-exchange separations can be performed only on ionic species. Spend a few minutes listing typical species that could be separated using ion-exchange methods.

We have listed a few typical compounds that have been the subject of ion-exchange separations:

>   inorganic cations and anions
>   amino acids
>   proteins
>   phenols
>   vitamins
>   amines
>   organic acids and bases

This list is not exhaustive and you may have thought of others. You may argue that molecules such as amines are not charged; don't forget that in solutions of low pH, amines can become protonated and hence positively charged. Conversely at high pH, acids can ionise, producing anionic species. These ions can then be separated by ion-exchange methods.

The fact that the above classes of compounds lend themselves to ion-exchange chromatography suggest there are large numbers of possible applications of ion-exchange. Bear in mind that many biological materials are based on amino acids and proteins and many drugs are based on similar materials or are organic acids or bases. Thus, in addition to the more common separations of inorganic species, ion-exchange methods have become important in a wide range of applications in biochemical, effluent and food analyses.

In this section we are going to look at only some of these areas. It is recommended that you do some of your own research in the chemical literature and in suppliers catalogues for those specific applications of the methods in which you have an interest.

5.3.1. Deionisation

This is the most often quoted application of ion-exchange, since it is used in practically all laboratories as a fast and effective method of water purification. Complete deionisation of water or a non-electrolyte solution is performed by exchanging solute cations for hydrogen ions and solute anions, for hydroxyl ions. Commercial production of deionised water is performed using alternating beds of cation and anion-exchangers. For purposes that require the highest purity, a mixed bed system is used in which the hydrogen form cation-exchanger and the hydroxyl form anion-exchanger are mixed together in a single column.

You may wish to examine the ion-exchange material that performs water purification in your laboratory. You will find that the resins used are based on styrene-divinylbenzene polymers. Notice that the presence of ions is detected by measuring the conductivity of the effluent from the ion-exchange column(s). This idea has been developed in the form of a highly efficient conductivity detector for high performance ion chromatography (Section 5.3.6).

5.3.2. Ultra-purification of Reagents

For many biological applications, there is a requirement to free reagents from contamination by metal ions. Buffer solutions prepared from reagent grade materials can be purified by passage over a suitable ion-exchanger.

One application of this method is the purification of urea. Fig. 5.3a illustrates the high efficiency of the column process as compared with the batch process.

	Batch	Column
Sample	100 cm^3 $6M$ urea	100 cm^3 $6M$ urea
Initial Conductivity	$70 \times 10^{-4} \text{ S m}^{-1}$	$70 \times 10^{-4} \text{ S m}^{-1}$
Amount of exchanger	5 g	5 g
Final Conductivity	$5 \times 10^{-4} \text{ S m}^{-1}$	$0.2 \times 10^{-4} \text{ S m}^{-1}$
Time taken	5 hours	10 minutes

Fig. 5.3a.   *Purification of urea by ion-exchange using both batch and column methods*

5.3.3.   Concentration of Trace Metals

Consider the situation where a sample of effluent contains lead present at a concentration below the limit of detection of analytical techniques such as atomic emission or absorption spectrometry.

Π   Suggest how ion-exchange could be used to increase the concentration of lead to a level where atomic spectrometry could be used in its detection.

Trace metals may be concentrated by retaining them on a strongly acidic cation-exchanger or a chelating resin. A known volume of the effluent is passed over the resin. The lead ions are held on the column where their effective concentration is increased. Sufficient effluent is passed over the column to raise the lead concentration to above that of the instrumental limit of detection. The retained lead can be subsequently released by treating the resin with a strong nitric acid solution; the hydrogen ions will displace the lead. Recoveries of 98% are common by this method and the column effluent will have a far higher concentration of lead than the original effluent sample, enabling quantitation by a technique such as atomic absorption spectrometry.

A modification of this type of application can be found in metal extraction. A manufacturing plant for nickel often produces effluent that contains low, but significant quantities of precious metals. Passing such an effluent through a cation-exchanger or a chelating resin often allows extraction of low grade effluents to become commercially viable.

There is an additional application for ion-exchange based on this particular idea. Can you suggest what the application is?

Many plants involved in metal production can present an environmental hazard by discharging metal salts in their waste water. Ion-exchange methods have been used extensively in effluent plants as a means of water purification prior to discharge to rivers.

5.3.4. Separation of Metals

One of the early applications of ion-exchange involved the separation of rare earth metals. Hydrochloric acid forms anionic complexes with many metals but exceptions are the alkali, alkaline earths, aluminium, nickel and chromium. Anion-exchangers can be used to separate these anionic chloro complexes because of the differences in the affinity of their complexes for the anion-exchanger as a function of the concentration of hydrochloric acid. Fig. 5.3b shows how the distribution ratios of metals change as a function of the hydrochloric acid concentration. Other complexing agents have also been used to effect the separation of metal ions.

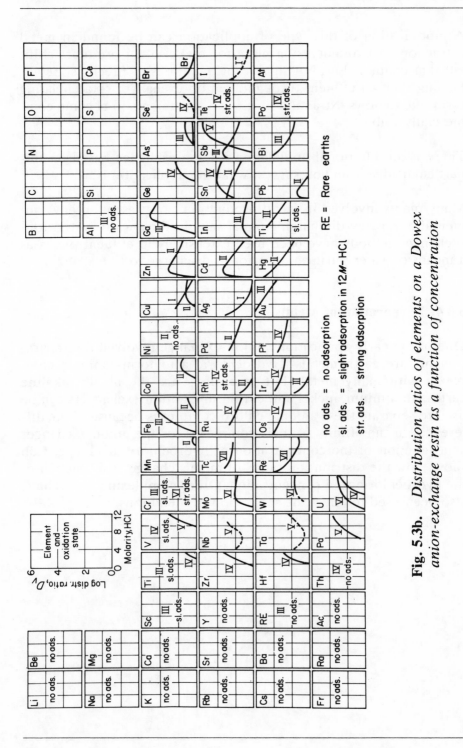

Fig. 5.3b. *Distribution ratios of elements on a Dowex anion-exchange resin as a function of concentration*

∏   Using the information obtained in Fig. 5.3b establish how a
     solution containing a mixture of nickel(II), copper(II) and
     zinc(II) ions could be separated by anion-exchange chro-
     matography.

The first point to note is that a high distribution ratio means a high
affinity for the anion-exchanger. Nickel ions do not exchange as they
form no chloro complex and are therefore eluted quickly from the
column. Comparison of the variation in the distribution ratio against
molarity of hydrochloric acid shows a significant difference in the
values for copper and zinc at molarities between 2 and 8. Thus, the
mixture could be separated by eluting with $2M$ hydrochloric acid
which will remove the copper from the column. However, the zinc
will be quite strongly held at this molarity. It will may be necessary
to lower the molarity of the hydrochloric acid further after elution
of the copper, so that the zinc ions elute from the column. This can
be done quickly by switching to a nitric acid mobile phase.

5.3.5.   Removal of Interfering Anions

In techniques such as atomic absorption spectrometry, anionic
species such as phosphate, sulphate, aluminate and other oxy-
genated anions depress the absorption of alkaline earth metals in
air/ethyne flames. The effect is particularly important when determining calcium and magnesium. Ion-exchange can be used to coun-
teract this type of interference. You may wish to apply your own
schemes of separation based on ion-exchange to a particular problem.

5.3.6.   Ion Chromatography

It is appropriate at this stage to mention ion chromatography, a
rapidly developing field of analysis. Although many of the applications already mentioned are performed on large columns, the development of ion-exchange resins with particle sizes of the order 10 μm
or less has enabled the production of highly efficient ion-exchange
columns. The availability of chromatographic detectors that oper-

ate on the principle of conductivity has introduced a new generation of applications. The technique extends the advantages of high performance liquid chromatography to the analysis and separation of inorganic ions in particular.

Examination of the current literature indicates a widespread application of the technique in fields as diverse as the food and beverage, electronics and pharmaceutical industries, in addition to environmental and clinical chemistry. There follows a brief summary of the analytical applications of ion chromatography; it is not meant to be a definitive list.

The Dionex Corporation was the first to market a commercial ion chromatography system. Their literature lists the ions and their detection limits separated by the system. This information is summarised in Fig. 5.3c. The same firm publishes over 50 notes covering specific applications such as the routine determination of chromate, the analysis of acids in pickling baths and the determination of trace ions in other matrices.

IONS ANALYZED AND MEASUREMENT SENSITIVITY

		Minimum Detection Limit		
		Sub μg/l (sub ppb)	μg/l (ppb)	mg/l (ppm)
COMMON ANIONS	Halides (F^-, Cl^-, Br^-)	•		
	Nitrate, Nitrite	•		
	Sulfate, Sulfite (Including combusted sulfur compounds)	•		
	Phosphates	•		
	Sulfide		•	
	Cyanide		•	
	Borate			•
COMMON CATIONS	Alkali Metals (Na, $^+$ K^+, Li^+)	•		
	Alkali earths (Mg^{2+}, Ca^{2+}, Sr^{2+}, Ba^{2+})	•		
	Ammonia		•	
	Hydrazine			•
ORGANIC ACIDS	Glycolic		•	
	Formic		•	
	Acetic		•	
	Propionic		•	
	Adipic		•	
	Glutaric		•	
	Succinic		•	
	Lactic		•	
	Malonic		•	
	Valeric			•
	Butyric			•

Fig. 5.3c. *Range of ions analysed and their detection limits – Dionex Ion Chromatograph*

Recently, a number of other firms have entered the market in ion chromatography, using conventional high performance liquid chromatographs modified with conductivity detectors and ion-exchange columns. Fig. 5.3d shows the simultaneous analysis of chloride, nitrate and sulphate in acid rain. Such a separation illustrates the advantage of ion chromatography over conventional wet chemical methods. A typical multi anion or cation analysis can be achieved by ion chromatography in minutes.

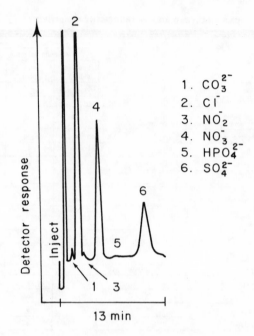

Fig. 5.3d. *Fast separation of anionic compounds in acid rain by ion chromatography*

Sample:	200 μl
Column:	1C-PAK^{TM}A
Eluent:	Borate Buffer
Flow Rate:	1 cm^3 min^{-1}
Detector:	Waters 430 conductivity, 1μSFS

Increasing use of ion chromatography has been found in clinical and pharmaceutical analysis. Fig. 5.3e shows the separation of benzodiazepines, a class of drug commonly used as a hypnotic and sedative. The separation is performed on a 10μm cation-exchange resin.

Ion-exchangers have also been used to separate amino acids and macromolecular species such as proteins. However, the latter are a class that can also be separated by size exclusion chromatography, a subject that will be dealt with in the next section.

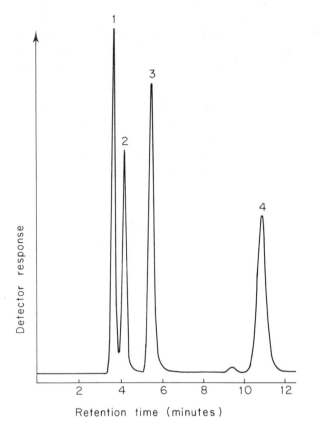

Fig. 5.3e. *Separation of benzodiazepines: (1) oxazepam, (2) ni-trazepam, (3) diazepam and (4) chlordiazepoxide. Column 25 × 0.46cm Partisil-10-SCX (bonded-phase 10-μm porous silica); mobile phase, 0.05 M ammonium phosphate (pH 3.0) in 60%v/v methanol/water; 22 °C; 1.0 ml/min, ΔP ≃ 350 psi; detector: UV absorption, 254 nm; 2 μl sample*

In the example given in Fig. 5.3e, the detector used was a uv absorbance monitor rather than a conductivity monitor. There is now an increasing interest in the use of reverse phase hplc on a non-polar column and a uv absorption dectector as an alternative to the Dionex system for the separation of both anions and cations. However, the mechanism of separation will not be discussed in this Unit.

SAQ 5.3a

(*i*) Iron(III) salts are often contaminated with iron(II). Suggest a scheme for the separation of these two ions by ion-exchange.

(*ii*) The determination of fluoride is subject to a number of interferences by metal ions. Suggest a scheme to separate metal anions from fluoride, and select an appropriate resin from a catalogue of ion-exchange materials.

(*iii*) Outline a method for detecting ions separated by ion chromatography in the absence of a conductivity detector.

5.3.7. Size Exclusion Chromatography

Earlier in this Unit you learnt that separations based on the mechanism of size exclusion allow molecules to be separated on the basis of molecular size and shape. Separation media are available that will separate macromolecular species with relative molecular masses from 400 to almost 10^{10}. Size exclusion chromatography is particularly suited to separations of natural and manufactured polymeric materials. The following classes of compound are typical:

> proteins
> polysaccharides
> resins
> adhesives
> polyesters

Let us see how the information gained from size exclusion work can be used in analysis.

5.3.8. Determination of Relative Molecular Mass

Size exclusion chromatography can be used to compare the relative molecular masses of polymeric materials. Calibration curves can be constructed by plotting the log of the relative molecular mass against elution volume. For protein determinations, the calibrants used are proteins of known relative molecular mass. Try this next example to understand how the technique is applied.

∏   The following data were obtained for a series of molecular mass 'markers'. Plot the log of the relative molecular mass against elution volume and hence determine the approximate relative molecular mass of gamma-globulin.

Marker	Approximate Relative Molecular Mass	Elution Volume cm^3
Thyroglobulin	670 000	12.5
Apo-ferritin	470 000	13.5
Transferrin	74 000	20.4
Haemoglobin	64 500	21.0
alpha-chymotrypsinogen	24 500	24.5
Cytochrome-c	13 000	26.4
gamma-globulin		18.3

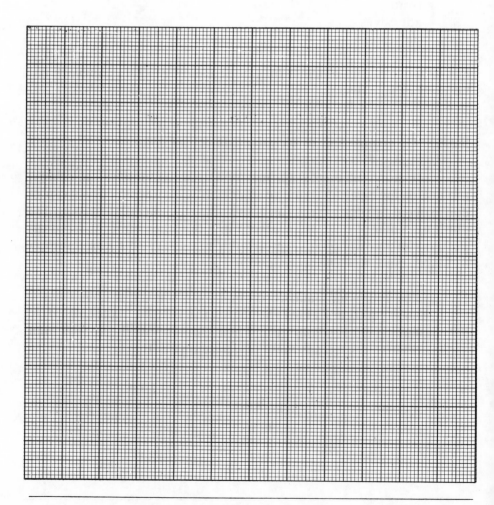

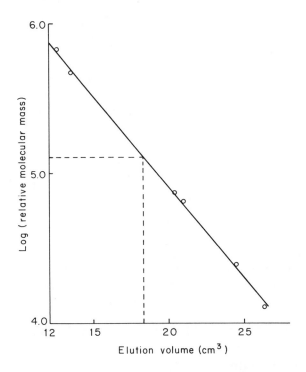

Fig. 5.3f. *Plot of log relative molecular mass against elution volume for a series of proteins*

The plot is shown in Fig. 5.3f. You will notice that there is an approximately linear relation between the log of the relative molecular mass and the elution volume. This is because the relative molecular masses of the proteins all lie within the range of selective permeation of the stationary phase. Provided that the unknown has similar characteristics to the marker materials, then an estimate of the relative molecular mass of the unknown can be made. In this case, you should have found the value for the approximate relative molecular mass of gamma-globulin to be 130 000.

Of course, this technique can be developed as a method of qualitative analysis for proteins by comparison of the retention volumes of particular types of protein. A typical protein analysis is shown in Fig. 5.3g.

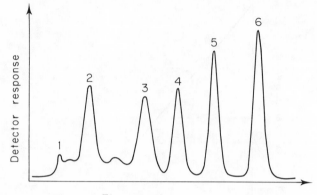

Fig. 5.3g. *Separation of a Bio-Rad size exclusion standard applied to a 1.5 × 71 cm column of Bio-Gel A-1.5m, 200–400 mesh. Flow rate 2.5 cm³ hr⁻¹*

Buffer 50 mM Tris, 145 mM NaCl, 4 mM EDTA, pH 7.6.
1. high molecular mass aggregates
2. thyroglobulin
3. gammaglobulin
4. ovalbumin
5. myoglobin
6. cyanocobalamin (vitamin B12)

This method of determining relative molecular mass is not limited to proteins. It can be applied to other polymeric systems, eg the determination of the molecular masses of polystyrene samples, provided that polystyrene molecular mass standards or 'markers' are used. The technique is thus useful in determining the 'average' relative molecular mass of a polymer.

5.3.9. Molecular Mass Distribution of Polymers

Characterisation of polymers is a difficult area of analysis that size exclusion methods can simplify. Manufactured polymers do not possess chains of identical length, but consist of a mixture of chains that can vary widely in molecular mass. The properties of polymeric systems depend intimately on the statistical distribution of these relative molecular masses. Size exclusion chromatography can be applied to determine many of the parameters used to define such systems. Properties such as molecular mass distribution, mass average and number average molecular masses can all be measured.

Size exclusion methods are thus widely used for quality control purposes and can be developed to recognise changes in polymer composition as a result of ageing, decomposition or mechanical damage.

Consider, for example, the changes in the condition of lubricating oils. With extended use, the polymer chains tend to break down and recombine to form compounds of high relative molecular mass. Monitoring the condition of the oil using size exclusion methods allows optimum efficiencies to be achieved for machinery.

Fig. 5.3h shows how the molecular masses of the lubricating oil chains change with time.

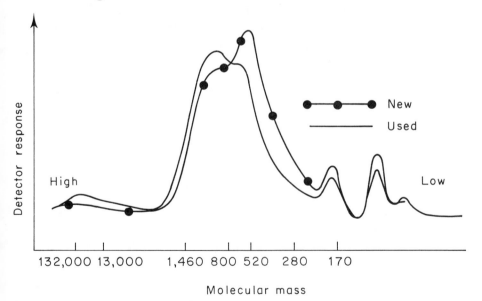

Fig. 5.3h. *Analysis of lubricating oil by size exclusion chromatography. The loss of lubricating properties results from an increase in the high molecular mass components in the oil*

5.3.10. Purification

In addition to the above analytical applications, size exclusion methods provide a means of preparing small and large quantities of polymeric samples in high states of purity. Certain drugs, serum proteins, amino acids peptides, RNA molecules and glycolipids can all be separated from impurity molecules by size exclusion chromatography. The scale of the preparation can be altered by changing column size and quantities of stationary phase. Since size exclusion does not depend on strong chemical interaction with the stationary phase, there is less danger of introducing chemical changes in the samples purified by this method. It is likely that this particular application will develop rapidly, particularly in areas associated with biotechnology.

SAQ 5.3b

(*i*) Compare the primary separation mechanisms of ion-exchange and size exclusion chromatography.

(*ii*) Define the terms *total exclusion limit* and *total permeation limit* for a size exclusion stationary phase.

Congratulations! You have now completed the text of this Unit. We hope you have completed all the SAQ's successfully; don't forget that the answers are contained at the end of the text. It is also well worth going through the check lists of Learning Objectives to confirm that you have achieved all of them.

We hope that you have enjoyed completing this Unit and hope that it will serve you as you progress to the other ACOL Units in chromatography. All the best!

Learning Objectives

You should now be able to:

● identify typical compounds separated by ion-exchange and size exclusion chromatography;

● list typical applications of the techniques in water purification, analysis and separation;

● explain how certain metals can be separated by anion-exchange chromatography of their complexes;

● describe briefly the technique of ion chromatography;

● state the advantages of ion chromatography over classical analysis;

● state two methods for the detection of ions separated by ion chromatography;

● state how size exclusion chromatography can be used in qualitative analysis;

● identify the use of size exclusion analysis in the analysis of polymeric materials.

Self Assessment
Questions and Responses

SAQ 1.2a   Complete the following diagrams showing how the components A, B and C of a mixture will appear as they pass down the chromatographic column in (*i*) elution development (*ii*) frontal analysis and (*iii*) displacement development. The affinity of the components for the column is in the order A < B < C.

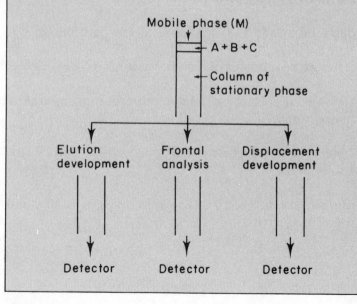

Response

In elution development the pure components will be separated by regions of mobile phase (M). Since the rate of migration is dependent on the affinity between the sample and the column, component A (with the smallest affinity) will have the highest rate of migration and will be eluted first. Component C will be the last to be eluted.

In frontal analysis, only the first component to be eluted (A) will be pure. Subsequent bands will consist of A and B and then A, B and C. Since the sample is being continually placed on the column, there will be no regions between the component bands where only pure mobile phase is being eluted and the third band will consist of A + B + C until all the sample has been used, when pure mobile phase will pass through the column.

In displacement analysis, the mobile phase displacer (D) will displace the component which is most strongly retained (C) which in turn acts as displacer for (B) which then displaces (A). There will be a region of overlap between the pairs (A + B) and (B + C).

Thus your diagram should show:

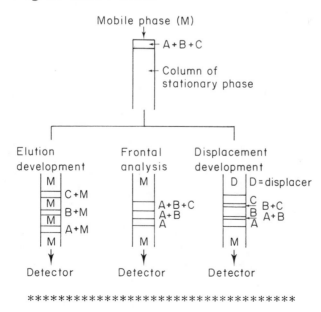

****

SAQ 1.2b

Following on from SAQ 1.2a, if a detector whose response is a function of component concentration is used to detect the eluent, complete the following diagrams to show how the signal will vary as the components A, B and C are eluted.

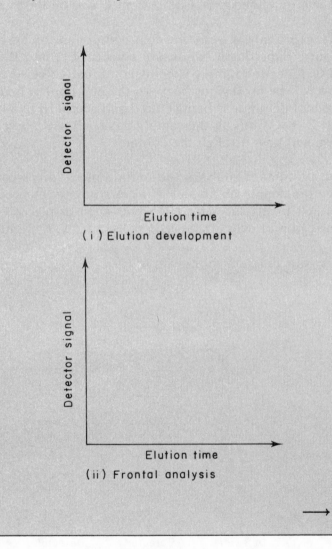

SAQ 1.2b (cont.)

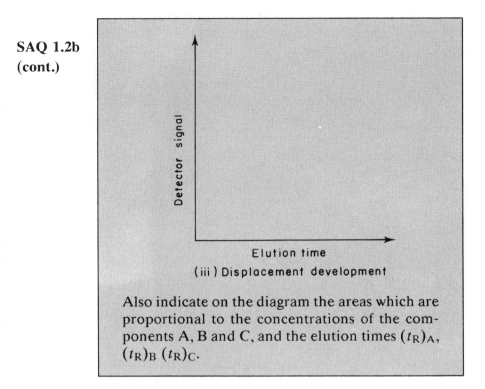

(iii) Displacement development

Also indicate on the diagram the areas which are proportional to the concentrations of the components A, B and C, and the elution times $(t_R)_A$, $(t_R)_B$ $(t_R)_C$.

Response

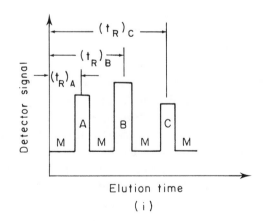

(i)

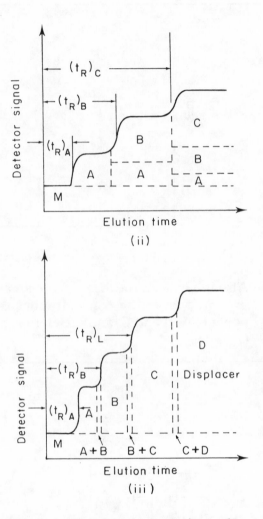

(ii)

(iii)

(*i*)   In elution analysis, the detector monitors the mobile phase as it elutes from the column to give the base-line. When component A passes through the detector the detector, signal will increase by an amount proportional to the concentration of A. After a time equivalent to the band width of component A, the detector signal will decrease to the base-line signal showing that A has passed through the detector. Thus, the area of peak A is proportional to the concentration of component A. The time taken for A to elute from the column measured from the time of injection ($t = 0$) to the centre of the peak A is the retention time of A. The signal will remain at the base-line level until the next component B is eluted when again a peak will be obtained, whose area is proportional to the concentration of B, and so on.

In practice, due to processes which occur in the column, the peaks are not plug-shaped as shown but are rounded off to give a Gaussian distribution.

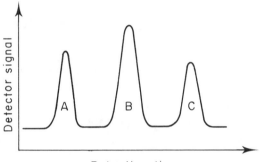

(*ii*) Frontal analysis. The recorder measures the base-line signal of the mobile phase until the least strongly retained component A breaks through at retention time t_A. The signal increases to a height which is proportional to the concentration of component A. However, in elution analysis, A on the column is being replenished from the sample reservoir so that the signal will remain constant and will not fall as in elution analysis. When the next component B elutes from the column after time t_B, the signal will again increase, but this time the height of the detector signal is due to A + B as shown. When C elutes from the column, the detector signal again increases and its height is now due to A + B + C. In principle, it is possible to calculate the proportions of each component in the original mixture, but in order to do so preliminary calibration is necessary and it is practical only for simple mixtures.

(*iii*) In the case of displacement development, the least strongly held component A is displaced first and the signal height after t_A is proportional to the concentration of A. Components B and C are eluted subsequently at times t_B and t_C and the heights of the steps again are proportional to the concentrations of B and C. The regions between the bands are the regions of overlap where the pure components are not obtained. After C has been eluted, the displacer itself breaks through.

****

SAQ 1.3a

The following results were obtained for the equilibrium concentrations of 2,4-pentanedione after extraction by tetrachloromethane from water.

Extraction	[2,4-pentanedione] in CCl$_4$/mol dm^{-3}	[2,4-pentanedione] in water/mol dm^{-3}
1	0.727	0.360
2	0.842	0.421
3	1.010	0.515

Calculate the distribution coefficient for 2,4-pentanedione between CCl$_4$ and water.

Response

The distribution coefficient is:

$$K_D = \frac{[\text{2,4-pentanedione}]_{\text{CCl}_4}}{[\text{2,4-pentanedione}]\,H_2O}$$

From Extraction 1: $K_D = 0.727/0.360 = 2.02$

Extraction 2: $K_D = 0.842/0.421 = 2.00$

Extraction 3: $K_D = 1.010/0.515 = 1.96$

The average value $K_D \dfrac{(CCl_4)}{H_2O} = 1.994$

Note that K_D is essentially constant although different amounts of 2,4-pentanedione were taken for the three extractions.

****

SAQ 1.3b

> 3 g of iodine is dissolved in a mixture of water and benzene (100 cm^3 of each) On analysis, 1 g of iodine is found in the water layer. If 5 g of iodine were dissolved in a mixture of water and benzene (100 cm^3 of each) at the same temperature, how much iodine would be found in the water layer?

Response

The distribution coefficient, K_D, is given by the expression

$$K_D = \frac{[I_2]_{Bz}}{[I_2]_{H_2O}}$$

From the first extraction, since 1 g of iodine is found in the water layer, the amount of iodine in the benzene layer must be $(3-1) = 2$ g.

$$\therefore \qquad K_D = \frac{(3.0 - 1.0)}{1.0} = 2.0$$

Since the distribution coefficient is a constant, we may use this value of K_D for the second experiment.

$$\therefore \qquad 2.0 = \frac{5 - x}{x}$$

where x is the amount of iodine dissolved in 100 cm^3 of water.

$\therefore$ $x = 1.67$g iodine in 100 cm^3 H$_2$O

****

SAQ 1.5a

Which mode(s) of chromatography would you use to separate the following mixtures?

(i)   inorganic gases, eg CO_2, CO, O_2 and SO_2

(ii)   polyaromatic hydrocarbons, eg anthracene, phenanthrene, pyrene and chrysene

(iii)   fatty acids from an oil, eg palmitic and oleic

(iv)   organo-chlorine pesticides, eg 2,4-D and 2,4,5-T

(v)   a mixture of Cl^-, SO_4^{2-} and CO_3^{2-} ions

(vi)   a mixture of Ca^{2+}, Sr^{2+} and Ba^{2+} ions

(vii)   saturated and unsaturated aliphatic hydrocarbons

($viii$) napthalene sulphonic acids

(ix)   a mixture of steroids

(x)   oligomers of an epoxy resin

(xi)   a mixture of amino acids

(xii)   a mixture of 1,2- 1,3- and 1,4-dihydroxy-benzenes

Response

(*i*)   Adsorption chromatography – since the sample components are already gaseous, they must be separated by gas chromatography. Their very low solubility in organic liquids means that they must be separated using the adsorption mode, ie using gsc.

(*ii*)   These compounds could be separated using glc, but the preferred method is now reverse phase liquid chromatography using a bonded stationary phase.

(*iii*)   Glc is the preferred method. However, long chain fatty acids such as these have high boiling points. Hence, they are derivatised to form the methyl esters to increase their volatility before separation using partition chromatography (see Section 2.3.3).

(*iv*)   Organo-chlorine pesticide are usually analysed by glc because of the availability of a detector capable of detecting them at the picogram level. They can be separated by hplc using adsorption chromatography, but the detection methods are less sensitive.

(*v*)   Anions may be separated by ion-exchange chromatography or by ion-pair chromatography.

(*vi*)   Cations are usually separated by ion-exchange, although ion-pair chromatography is being used more frequently.

(*vii*)   GLc is the best mode to use for separating members of an homologous series. This could be done by reverse phase hplc but the detectors used in hplc are not very sensitive for this type of molecule.

(*viii*) Ionic or ionisable organic compounds like the napthalene sulphonic acids can be separated by ion-pair chromatography.

(*ix*)   Steroids, like long chain fatty acids, are very polar (due to the $-OH$ groups) and have a low volatility. They can be separated by glc as the trimethysilyl derivatives or by reversed phase hplc.

(*x*) Epoxy resin oligomers would be too involatile to separate by gc. and are therefore separated by exclusion chromatography.

(*xi*) Amino acids were for many years separated by ion-exchange chromatography using the ninhydrin reaction to detect them, but more often today they are analysed using reverse phase liquid chromatography and a uv absorbance detector.

(*xii*) We saw that adsorption chromatography was a good method for separating geometric isomers because of selective adsorption (Section 1.5.1).

****

SAQ 2.1a

> The retention times of peaks A, B and C, where A is an unretained component, are 0.84 min, 10.60 min and 11.08 min respectively. If the volume of stationary phase is 12.3 cm^3 and the mobile phase flow rate is 20.95 cm^3 min^{-1} calculate:
>
> (*i*) The retardation factors for B and C and
>
> (*ii*) the distribution coefficients of B and C.

Response

(*i*) The retardation factor is given by:

$$= \bar{u}_x/\bar{u}_m \tag{2.1}$$

or for a column of length L and retention times t_R and t_m,

$$R = \frac{L/(t_R)_x}{L/t_m} = \frac{t_m}{(t_R)_x}$$

$$\therefore \qquad R_A = \frac{0.84}{10.60} = 0.079$$

and $$R_B = \frac{0.84}{11.08} = 0.076$$

(*ii*) The distribution coefficient is obtained by rearranging Eq. 2.14

$$V_R = V_m + KV_s$$

or $$K = (V_R - V_m)/V_s$$

Therefore to calculate K we require the retention volumes $(V_R)_B$, $(V_R)_C$ and V_m.

where $(V_R) = F \times t_R$

$$\therefore \qquad (V_R)_B = 20.95 \times 10.60 = 222.07 \text{ cm}^3$$

$$(V_R)_C = 20.95 \times 11.08 = 232.13 \text{ cm}^3$$

and $$V_m = 20.95 \times 0.84 = 17.60 \text{ cm}^3$$

$$\therefore \qquad K_C = \frac{222.07 - 17.60}{12.30} = 16.62$$

and $$K_C = \frac{232.13 - 17.60}{12.30} = 17.44$$

$\therefore$ Retardation factor of A = 0.079
Distribution coefficient of A = 16.62

Retardation factor of B = 0.076
Distribution coefficient of B = 17.44

****

SAQ 2.1b   From the given chromatogram calculate for peak B:

(*i*)   the adjusted retention volume

(*ii*)   the net retention volume

(*iii*) the specific retention volume

The mobile phase flow rate is 40.0 cm^3 min^{-1}, the chart speed = 0.5 cm min^{-1}, column inlet pressure (P_i) = 1.2 bar column outlet pressure (P_i) = 1.0 bar, column temperature = 100 °C and weight of stationary phase = 1.2 g. Peak A has $K = 0$.

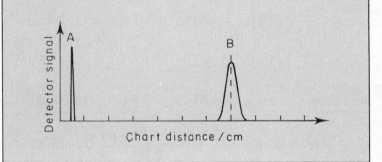

Response

(*i*)   Adjusted retention volume V'_R

To calculate the adjusted retention volume we need the retention time (t_m) of the unretained peak (A) and the uncorrected retention time (t_R) of peak B.

Both these values are measured from the point of injection (ie the start of the chromatogram) in units of 'chart distance' and then converted to units of time.

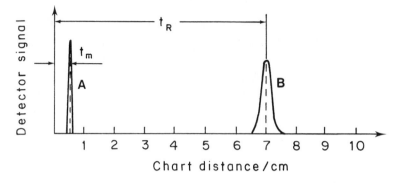

Thus, $t_m = 0.5$ cm and $t_R = 7.0$ cm

or, since the chart speed $= 0.5$ cm min^{-1},

$$t_m = \frac{0.5}{0.5} = 1.0 \text{ min and } t_r = \frac{7.0}{0.5} = 14.0 \text{ min}$$

∴   the adjusted retention volume,

$$V_R' = F \times t_R' = F(t_R - t_m)$$

$$= 40.0 \times (14.0 - 1.0) = 520 \text{ cm}^3$$

(ii)   Net Retention Volume, V_N

This is the adjusted retention volume corrected for the mobile phase compressibility ie $V_N = jV_R'$

$$j = \frac{3}{2}\left[\frac{(P_i/P_o)^2 - 1}{(P_i/P_o)^3 - 1}\right] = \frac{3/2}{2}\left[\frac{(1.3/1.0)^2 - 1}{(1.2/1.0)^3}\right]$$

$$= \frac{3}{2}\left[\frac{1.44 - 1}{1.73 - 1}\right] = \frac{3}{2}\left[\frac{0.44}{0.73}\right] = 0.90$$

Net retention volume $= jV_R' = 0.90 \times 520$

$$= 468.0 \text{ cm}^3$$

(*iii*) Specific retention volume V_g

This is the net retention volume corrected to 0 °C (273.16K) per gram of stationary phase.

$$V_g = \frac{V_n 273.16}{W_s T_c}$$

$$= \frac{468.0 \times 273.16}{12 \times 373.16} = 285.5$$

$\therefore$ Specific retention volume $= 285.5 \ cm^3$

****

SAQ 2.3a

> Which of the following molecules would you expect to be *polar*? Pauling electronegativity values are C (2.5), H (2.1) and Cl (2.9).
>
> (*i*)   benzene C_6H_6
>
> (*ii*)   chlorobenzene C_6H_5Cl
>
> (*iii*)   *n*-butane $CH_3-CH_2-CH_2-CH_3$
>
> (*iv*)   cyclohexane C_6H_{12}

Response

(*i*)   If you think benzene is polar you are wrong. Although there is an electronegativity difference between C and H the molecule as a whole is symmetrical and can be drawn:

Thus the molecule is symmetrical and there is no net movement of electrons to any atom or group of atoms.

(*ii*) Chlorobenzene is *polar*. The strongly electronegative chlorine atom attracts electrons and produces a partial charge separation

$$\overset{\delta+}{C} - \overset{\delta-}{Cl}$$

This separation of charge gives rise to a dipole moment which can be used as a measure of polarity.

(*iii*) and (*iv*) Neither butane nor cyclohexane are polar since, like benzene, they are symmetrical molecules.

****

SAQ 2.3b

Which of the following molecules could experience an induced dipole?

(*i*)   *n*-butane

(*ii*) buta-1,3-diene

(*iii*) cyclohexane

(*iv*) benzene

Response

n-Butane and cyclohexane do not contain a conjugated electron system (eg —C=C—C=C—etc) and are not polarisable, buta-1,3-diene and benzene are conjugated and the electrons are not re-strained between the atoms as is suggested by the conventional way of drawing the structures of these molecules. The electrons are said to be *delocalised* and we can draw the structures as shown below.

Butadiene $H_2C \bar{\ } CH \bar{\ } CH \bar{\ } CH_2$

Benzene

where the dotted line indicates the movement of delocalised electrons.

****

SAQ 2.3c   Explain the following:

(*i*)   *n*-pentane is insoluble in water but *n*-butyl alcohol and diethyl ether have about the same solubility (8 g in 100 cm^3 water).

(*ii*) The boiling points of diethyl ether, *n*-pentane and *n*-butyl alcohol are 35 °C, 36 °C and 117 °C respectively.

Response

(*i*)   Although the ether group cannot donate a hydrogen atom it can accept one so that ethers are soluble in water together with alcohols.

(*ii*) The ether molecules are not polar enough to attract each other any more than those of *n*-pentane and therefore have similar boiling points because of their similar relative molecular masses (C_5H_{12} = 72; $C_4H_{10}O$ = 74) *n*-butyl alcohol/(M_r = 74) is associated and therefore has a boiling point which reflects its higher relative molecular mass in the associated form.

****

SAQ 2.3d

The following two chromatograms were obtained on (*i*) squalane (a hydrocarbon) and (*ii*) polyethyleneglycol (PEG) (a substance with an ether-like structure

$$-CH_2-CH_2-O-CH_2-CH_2-O-CH_2-CH_2-OH).$$

The three peaks are:

		Boiling Point	Polarity
A.	methanol	65 °C	high
B.	methyl ethanoate	57 °C	medium
C.	diethyl ether	36 °C	low

Assign the components A, B and C to the three peaks in the two chromatograms, and give your reasoning.

$\longrightarrow$

**SAQ 2.3d
(cont.)**

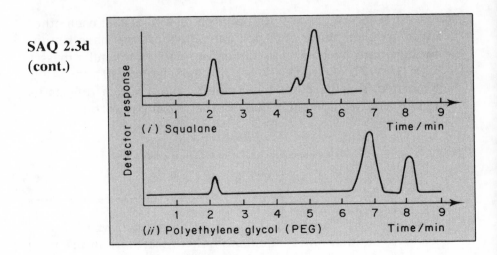

(i) Squalane

(ii) Polyethylene glycol (PEG)

Response

Since squalane is a hydrocarbon, it will be non-polar whereas polyethylene glycol will be polar due to its terminal —OH groups. Furthermore, the —O—groups in PEG will be available for hydrogen bonding.

We can assign the peaks on the squalane chromotagram as follows:

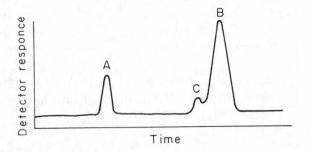

Peak A (methanol), although it has the highest boiling point (lowest vapour pressure) shows least affinity for squalane because of its high polarity. The assignment of peaks B and C is more difficult, and you may well have them wrong. Diethyl ether (C) is the less polar and would therefore have the highest affinity for squalane, but it also has a very high vapour pressure (low boiling point). Methyl ethanoate (B) has a lower affinity for squalane since it is more polar than diethyl ether, but it also has a low vapour pressure (higher boiling point). Thus with ether and methyl ethanoate on squalane there are two opposing effects (polarity and vapour pressure) and it is difficult to predict the actual order of elution. Certainly diethyl ether shows more retention than its boiling point would suggest, but only experiment would determine the actual order.

On the PEG chromatogram we can make the following assignments:

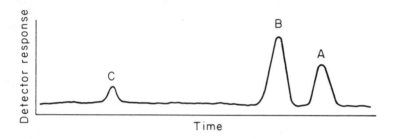

Here the assignment is more straightforward since the polarity affect is the controlling factor. Diethyl ether (C) is relatively non-polar, and that coupled with its high vapour pressure will mean that it will have a very short retention time.

Methanol (A), with its high polarity and boiling point (low vapour pressure), will have the longest retention time, and methyl ethanoate will be eluted in between methanol and diethyl ether.

****

SAQ 3.3a   (*i*) Calculate the asymmetry factor of the peak below:

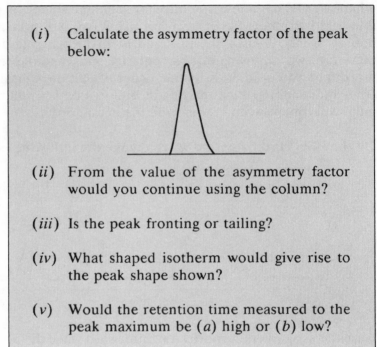

(*ii*) From the value of the asymmetry factor would you continue using the column?

(*iii*) Is the peak fronting or tailing?

(*iv*) What shaped isotherm would give rise to the peak shape shown?

(*v*)   Would the retention time measured to the peak maximum be (*a*) high or (*b*) low?

Response

(*i*)   The asymmetry factor is given by the ratio CB/AC at 10% of the peak height, ie $A_s = 6/5 = 1.2$.

(*ii*) The value of $A_s = 1.2$ indicates a poor quality column but it has not deteriorated to a point where it cannot still be used.

(*iii*) The peak is tailing ($A_s > 1$).

(*iv*) A Langmuir isotherm gives a tailing peak.

(*v*)   With a tailing peak, the zone moves more rapidly than a symmetrical (Gaussian) distribution so that the measured t'_R would be low (see Fig. 3.3l(i)).

****

SAQ 3.4a The following chromatogram was obtained from a 2 metre packed column. Calculate the values of N and H for the second peak using two different methods. Comment on any difference you get.

Note that the first peak is an unretained component.

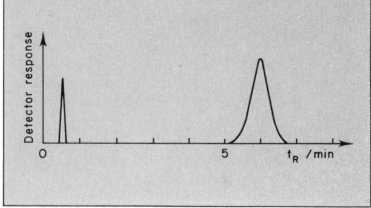

Response

The most suitable equations to use are 3.3 and 3.4.

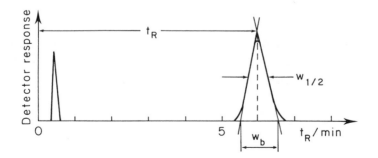

Using Eq. 3.3 $\quad t_R = 6.0 \text{ min}, \ w_{\frac{1}{2}} = 0.5 \text{ min}$

$$\therefore \qquad N = 5.54 \left(\frac{6.0}{0.55}\right)^2 = 659$$

$$\therefore \qquad H = \frac{2000}{659} = 3.03 \text{ min}$$

Using Eq. 3.4 $\quad t_R = 6.0 \text{ min}, w_b = 1.0 \text{ min}$

$$\therefore \qquad N = 16 \left(\frac{6.0}{1.0}\right)^2 = 576$$

$$\therefore \qquad H = \frac{2000}{576} = 3.47 \text{ min}$$

If you did not get these values, check that you have measured t_R correctly ie from the point of injection, not the corrected retention time measured from the unretained peak.

In this exercise it may be that the result from Eq. 3.4 is the more accurate because there is no difficulty in establishing the base-line. This may not be the case in practice.

$$*************************************$$

SAQ 3.4b   From the chromatogram in SAQ 3.4a calculate the values of N_{eff} and H_{eff} for the second peak.

Response

Using the values measured earlier and the value for $t_m = 0.5$ min

$$t_R = 6.0 \text{ min}$$

$$t_m = 0.5 \text{ min}$$

$$W_{\frac{1}{2}} = 0.55 \text{ min}$$

$$N_{eff} = \left(\frac{t_R - t_m}{W_{\frac{1}{2}}}\right)^2$$

$$= 5.54 \left(\frac{6.0 - 0.5}{0.55}\right)^2 = 554$$

$$H_{eff} = L/N_{eff} = 2000/554$$

$$= 3.61 \text{ mm}$$

****

SAQ 3.6a

SAQ 2.3d shows the separation of methanol, methyl ethanoate and diethyl ether on squalane and polyethylene glycol. Given that $t_m = 0.3$ min:

(*i*) calculate the separation factors for the component pairs: methyl ethanoate/methanol; methyl ethanoate/diethyl ether and diethyl ether/methanol on the two columns.

(*ii*) on the basis of the α values, which column would you use to separate each pair of compounds?

Response

(i) From the two chromatograms we get:

Retention times

		Column		
Solutes	Squalane		Polyethylene Glycol (PEG)	
	t_r	t'_r	t_r	t'_r
Methanol	2.1	1.8	7.9	7.6
Methyl ethanoate	5.1	4.8	6.7	6.4
Diethyl ether	4.6	4.3	2.2	1.9

Values of α

	Column	
Solute pairs	Squalane	PEG
Methyl ethanoate/methanol	2.67	0.84
Methyl ethanoate/diethyl ether	1.12	3.37
Diethyl ether/methanol	2.39	4.00

The better column for each individual pair of components will be that which gives the larger value of α, hence:

for methyl ethanoate/methanol – squalane

for methyl ethanoate/diethyl ether – PEG

for diethyl ether/methanol – PEG

From the value of α (0.84) for methyl ethanoate/methanol on PEG, we might think that the separation is poor since it is a lower value than for methyl ethanoate/diethyl ether on squalane (1.12). However, if we invert the ratio so that α is greater than unity, we have α (methanol/methyl ethanoate) = 1.19 and a slightly better separation than methyl ethanoate/diethyl ether on squalane (1.12).

****

SAQ 3.6b	Assuming that the two peaks can be represented by two triangles, calculate the value of R_s which corresponds to the situation where the two peaks are *just* separated at the base-line.

Response

The figure shows the two peaks just separated at the base-line (this is called base-line resolution).

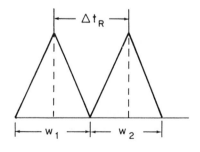

Then $R_s \qquad = \dfrac{t_{R_2} - t_{R_1}}{\frac{1}{2}(W_1 + W_2)} = \dfrac{\Delta t_R}{\frac{1}{2}(W_1 + W_2)}$

By inspection,

$$\Delta t_R = \frac{1}{2}(W_1 + W_2)$$

$$\therefore \qquad R_s = \frac{\frac{1}{2}(W_1 + W_2)}{\frac{1}{2}(W_1 + W_2)} = 1.0$$

Of course, you may not have drawn two equal triangles, since the question did not specify this, but the answer would have been the same as can be seen from the figure below, since again $\Delta t_R = \frac{1}{2}$ $(W_1 + W_2)$ and $R_s = 1.0$.

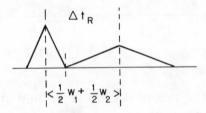

****

SAQ 3.6c The following two chromatograms represent the separation of a 2:1 mixture of cyclohexane and benzene on two different columns, column 1 being apolar and column 2 polar. For column 1, the chart speed was 1 cm s^{-1} and for column 2 it was 0.5 cm s^{-1}. All other chromatographic conditions were the same in each case.

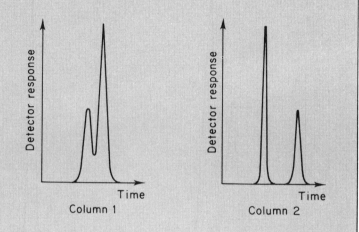

Column 1   Column 2

(*i*) Calculate the values of R_s for the two columns and comment on the results you get.

(*ii*) Assuming that the detector response is proportional to concentration, how do you account for the change in the elution order of cyclohexane and benzene on the two columns?

Response

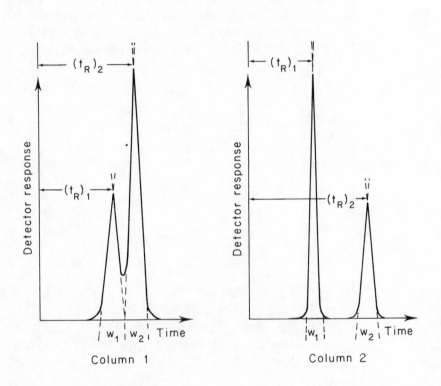

Column 1   Column 2

Drawing in the tangents to the peaks and the retention times from the injection point ($t = 0$) you should get the following values.

(*i*) Column(1) apolar Chart speed 1.0 cm min^{-1}

$(t_R)_1 \equiv 2.5$ cm $\equiv 2.5$ min. $W_1 = 0.8$ cm $\equiv 0.8$ min

$(t_R)_2 \equiv 3.3$ cm $\equiv 3.3$ min. $W_2 = 0.8$ cm $\equiv 0.8$ min

$$\therefore \quad R_s = \frac{3.3 - 2.5}{\frac{1}{2}(0.8 + 0.8)} = \frac{0.8}{0.8} = 1.0$$

This resolution is insufficient for base-line separation,

(*ii*) Column (2), polar, Chart speed $= 0.5$ cm min^{-1}

$(t_R)_1 \equiv 2.2$ cm $\equiv 4.4$ min. $W_1 \equiv 0.4$ cm $\equiv 0.8$ min

$(t_R)_2 \equiv 4.0$ cm $\equiv 8.0$ min. $W_2 \equiv 0.6$ cm $\equiv 1.2$ min

$$\therefore \quad R_s = \frac{8.0 - 4.4}{\frac{1}{2}(0.8 + 1.2)} = \frac{3.6}{1.0} = 3.6$$

Since $R_s > 1.5$, base line separation is achieved.

[Note. If you got an answer $R_s = 7.2$ in part (2), you probably did not convert the peak width measurement into units of time.]

(*ii*) the order of elution on the apolar column is determined by the volatility (ie boiling points) of the two components. The boiling points of cyclohexane (80.7 °C) and benzene (80.1 °C) are very similar. Benzene, the more volatile, is eluted first but the difference in volatility is too small to give complete separation. On the polar column, benzene with its delocalised π-electrons, has an additional interaction with the stationary phase in the form of dipole-induced dipole forces. Cyclohexane, which is not polarisable, has no such interaction. Therefore on the polar column, benzene is retained more than cyclohexane and the order of elution is reversed.

If you could not answer this part of the SAQ, refer back to Section 2.3.1. Here we saw that dipole-induced dipole interactions occur when a charge on one molecule can polarise the electrons in a second molecule. Thus although benzene is not polar, it is polarisable and on the principle of 'like has an affinity for like' (Section 2.3.3) there will be more interaction between benzene and the polar column than there is between cyclohexane and the polar column. Hence, benzene will be retained preferentially on the polar column and will have the longer retention time.

****

SAQ 3.6d From the standard resolution curves given, estimate the resolution of the two peaks in the chromatograms below.

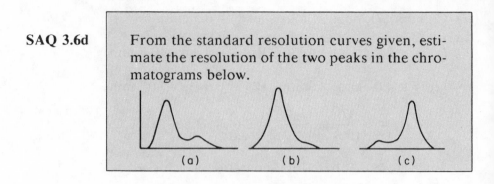

(a)   (b)   (c)

Response

(*a*) 1.0 (*b*) 0.7 (*c*) 1.0.

In example (*a*) the peak height ratios are clearly 4:1 so that Fig. 3.6d(*ii*) is the correct one to use. Inspection of the chromatograms shows that the best match is given with $R_s = 1.0$.

(*b*) When the peak band ratios are large it is not so easy to estimate their values. There is obvious tail-flattening at about a ratio of 16:1, so that Fig. 3.6d(*iv*) is the one to use. Deciding on the R_s value is again more difficult, but the best match is obtained with $R_s = 0.7$.

(*c*) In this example, the minor peak is eluted first. This does not prevent us using the standard curves, since a mirror image of our chromatogram can be superimposed on them. The band size ratio is 1:8, so that Fig. 3.6d(*iii*) is used, and the mirror image of our chromatogram matches best with the standard curve at $R_s = 1.0$.

****

SAQ 3.6e	For the chromatogram given below (*a*) determine the R_s value for the chromatogram and (*b*) calculate the R_s value for peaks 4 and 5 and the percent recovery at the equal-purity cut-point.

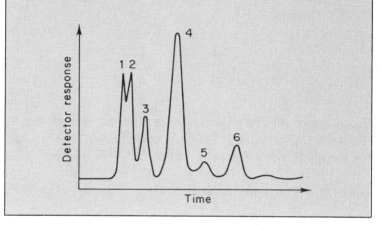

Response

(*a*) We saw that in defining the R_s value for a chromatogram it was usual to consider the least well resolved pair of peaks, so that R_s for any other pair would be at least equal to this.

Peaks (1) and (2) are the least well resolved. Their peak height ratio is $1:1$ so that we must look at Fig. 3.6d(*i*). Comparison with the standard curves suggests that the R_s value ≈ 0.7.

(*b*) For peaks (4) and (5), the peak height ratios are $8:1$ and $1:8$ respectively. Comparison with Fig. 3.6d(*iii*) gives a value R_s ≈ 1.0 and the equal polarity cut point is thus 98%. To obtain the recovery we must refer to Fig. 3.6h. Although the figure does not give a value for 8/1, we can estimate where its position would be:

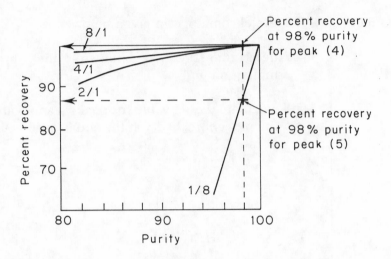

Looking to where a vertical line drawn through the 98% purity intersects the estimated 8/1 line and the 1/8 line we can read off the percent recovery for peak (4) as > 99%, and for peak (5) ≈ 86%.

Thus the complete answer is:

(*a*) Resolution of chromatogram > 0.7

(*b*) Resolution of peaks (4) and (5) = 1.0

Percent recovery of (4) at 98% purity > 99% Percent recovery of (5) at 98% purity ≈ 86%.

****

| SAQ 3.6f | Describe a general practical strategy for obtaining a satisfactory resolution in terms of k', N and α. |

Response

(*i*) Decide on the minimum acceptable value of R_s, eg 1.0.

(*ii*) Select a column to give values of $\alpha > 1.2$.

(*iii*) Run the chromatogram and estimate (or calculate) R_s for the least well resolved pair of peaks. Compare the value with the required value.

(*iv*) Optimise the value of N by using the flow rate corresponding to H_{min} etc.

(*v*) Adjust the chromatographic conditions to give $1 < k' < 10$ using temperature programming in gas chromatography and gradient elution in liquid chromatography.

(*vi*) If the required R_s is still too low in liquid chromatography, a change in the mobile phase composition should be considered.

****

SAQ 3.6g   The figure below shows different resolution problems (*a*) (*b*) and (*c*) requiring different separation strategies. Which of the three parameters k' N and α would you change to produce the desired resolution in each case?

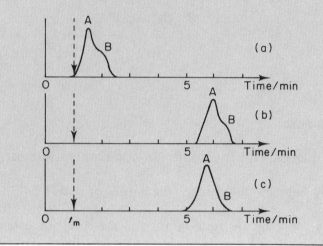

Response

(*a*) The retention times are too short ie both k' values are too low ($k'_2 < 1$) so that k'_2 must first be increased to bring it into the optimum range,

(*b*) Here k'_2 is in the optimum range ($k'_2 \approx 5$) but the resolution is insufficient. The best solution will be an increase in N.

(*c*) Although k' is again optimal ($k' \approx 5$), the resolution is very small and a very large increase in N would be required to produce a separation. This would result in very long retention times. It would be better to change the stationary phase and/or mobile phase to increase α in this case.

****

SAQ 3.6h

The figure below shows the H versus $\bar{u}$ curve obtained on a packed column in gas chromatography. Calculate the values of A, B and C in the van Deemter equation,

$$H = A + B + C\bar{u}$$

[It may be helpful to redraw the curve onto graph paper.]

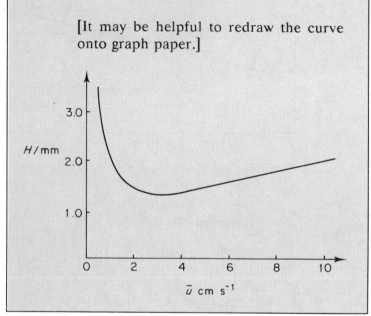

Response

The figure shows the above curve with the asymptote drawn in. Since the asymptote will be very close to the actual curve at high flow rates $(10\ \text{cm s}^{-1})$ the vertical distance (d) between asymptote and curve will be 0.01 mm.

$$\left(\text{since } d \propto \frac{1}{\bar{u}} = \frac{1}{100}\ \text{mm}\right)$$

At a flow rate of 5 cm s^{-1}, d will be 0.02 mm ($d = 1/50$ mm) etc.

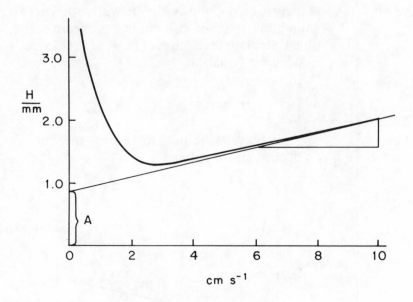

∴ A = intercept = 0.9 mm

           B = (10.0 × 0.1, 50 × 0.2 etc) = 10.0 mm^2 s^{-1}

           C = slope (0.46/40) = 0.011 s

further H_{min} = 1.3 mm

         $\bar{u}_{min}$ = 3.0 cm s^{-1}

These values are typical for a packed gas chromatography column.

****

SAQ 3.6i

From a consideration of the van Deemter equation predict the affect of the following changes on the value of the plate height H.

(*i*)   a decrease in particle size

(*ii*)   the use of a narrow particle size range

(*iii*) the use of spherical particles

(*iv*) an increase in the density of the mobile phase

(*v*)   an increase in mobile phase pressure

(*vi*) an increase in flow rate

(*vii*) a decrease in temperature

(*viii*) a decrease in the amount of stationary phase

(*ix*) a decrease in the viscosity of the stationary phase

Response

(*i*)   A reduction in particle size will decrease the multiple path effect and hence H, as long as the particles can still be packed uniformly, so that λ does not increase. Uniform packing is assisted by using spherical particles of a narrow size range, so that (*ii*) and (*iii*) also help in reducing H.

(*iv*) an increase in the mobile phase density will decrease the diffusivity of the solute species. At low flow rates, when the contribution from longitudinal molecular diffusion is large, this will lead to a decrease in H. However, at high flow rates, when mass transfer is more important, the increased density will reduce mass transfer in the gas phase. For fast analysis lighter gases, eg helium and hydrogen, are therefore preferred.

(*v*)   in gas chromatography, an increase in the mobile phase pressure produces a larger pressure drop across the column and this means that less of the column can be operated at the optimum flow rate. Therefore H will increase.

(*vi*)   the effect of a change in mobile phase velocity will depend on the position of the minimum in the van Deemter curve. Any change in flow rate away from this point will lead to an increase in H. The effect is far less, however, for changes on the high flow rate side of $\bar{u}_{min}$.

(*vii*)   the effect of a change in temperature is complex. Distribution coefficients usually increase with a decrease in temperature, so that the capacity factor and retention increases for flow velocities above $\bar{u}_{min}$. A decrease in temperature will also increase H because of the increasing mass transfer terms. Molecular diffusion (the B term) will decrease with a decrease in temperature.

(*viii*) a decrease in the amount of stationary phase will decrease k' and the film thickness d_f, and the overall effect will be a decrease in the C term and hence in H.

(*ix*)   a decrease in the stationary phase viscosity will lead to an increase in solute diffusion in the stationary phase (ie better mass transfer) which will again decrease C and hence H.

****

SAQ 3.6j	In a reverse phase system (using, say, a non-polar stationary phase, what would be the effect on the k' values of changing the mobile phase from methanol to acetonitrile?

Response

In a reverse phase system, the strongest solvent is the least polar, and since by definition, a strong solvent is one which elutes the sample rapidly, a change from methanol ($\epsilon^0 = 0.95$) to acetonitrile ($\epsilon^0 = 0.65$) would decrease the k' value. See Section 2.3.2 if you did not get this right.

****

SAQ 4.1a	A laboratory has been requested to analyse a mixture known to contain three aromatic hydrocarbons, the identities of which are suggested to be anthracene, pyrene and phenanthrene.
	The laboratory decides to perform a thin-layer analysis against standards of anthracene, pyrene and phenanthrene respectively that the laboratory has in stock. The resulting chromatogram shows that the three components have R_f values that correspond to the R_f values of the three standards. On this basis the laboratory reports that the identity of the mixture is confirmed.
	(*i*) Do you consider their report to be valid?
	(*ii*) State the reasons for your conclusion in (*i*)
	(*ii*) Suggest alternative means of confirming their result.

Response

Their result is certainly not valid, because of the vast number of
polynuclear aromatic compounds. It is possible that others may have
R_f values that correspond to those obtained in this single analysis.
The identities could be confirmed only by performing a spectrometric examination of the separated components, although less reliable
evidence could be obtained by repeating the analysis with a number
of different solvent combinations to discover whether the R_f values
of sample components and standards were identical in each case.

****

SAQ 4.1b

The corrected retention times t_R, in seconds
for the following series of compounds were
determined on a non-polar (apolar) column.
Calculate, using a graphical and non-graphical
method, the Kovats Indices for methylbenzene and cyclohexane relative to the series of
n-alkanes.

Compound	Corrected Retention Time t_R/s
n-butane	7.1
n-pentane	17.8
n-hexane	38.0
n-heptane	83.0
n-octane	182
methylbenzene	100
cyclohexane	79.0

Response

A Kovats plot for a series of *n*-alkanes would be as shown:

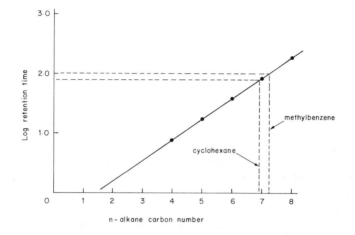

Kovats Index for methylbenzene $= 7.25 \times 100 = 725$
Kovats Index for cyclohexane $= 6.92 \times 100 = 692$

The retention index may be calculated from the following equation:

$$I = 100 \left[n \left\{ \frac{\log t_R(x) - \log t_R(y)}{\log t_R(x + n) - \log t_R(y)} \right\} + y \right]$$

where I is the Kovats Index,

$t_R(x)$ is the corrected retention time of the unknown *x*

$t_R(y)$ is the corrected retention of the *n*-alkane with carbon number *y*

$t_R(x + n)$ is the corrected retention time of the *n*-alkane with carbon number $(x + n)$

n is the difference in number of the two reference *n*-alkanes.

If *n*-heptane and *n*-octane are taken as the reference hydrocarbon, the following values may be substituted for the sample of methyl benzene.

$$\log t_R(x) = 2.00 \qquad \log t_R(y) = 1.919$$

$$\log t_R(x - n) = 2.26 \qquad n = 1 \qquad y = 7$$

$$I = 100 \left[1\left\{ \frac{2.000 - 1.919}{2.260 - 1.919} \right\} + 7 \right]$$

$$= 724$$

A similar calculation for cyclohexane using n-hexane and n-heptane as reference give a value of 693.

****

SAQ 4.2a   The following values were obtained from an electron capture detector when analysing standards of lindane, a pesticide:

Response (R) (arbitrary units)	log R	Mass of lindane injected $(M$/ng)	log M
0.56	−0.25	0.100	−1.00
2.01	0.303	0.316	−0.50
9.77	0.990	1.00	0.00
89.1	1.95	10.00	1.00
200	2.30	100	2.00
224	2.35	316	2.50

$\longrightarrow$

SAQ 4.2a
(cont.)

**SAQ 4.2a
(cont.)**

(*i*) Plot the log of the response against log *M*.

(*ii*) Comment on the shape of the curve, and give the range of linear response of the system and the lower limit of detection.

(*iii*) Two samples of lindane, extracted from a food sample, give readings of 40.3 and 400 for the detector response. Calculate the mass of lindane extracted from the sample, commenting on your results.

Response

(*i*) Your graph should be identical to that shown below:

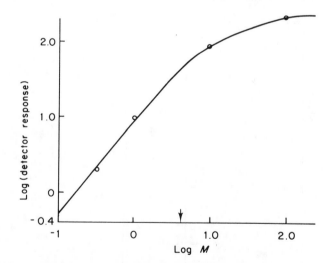

(*ii*) The shape of the curve has a linear portion, but over a much lower range of response and sample concentration than the flame ionisation detector discussed previously. The linear response will limit the range of concentration of lindane studied to between 0.1 and 6.3 ng.

(*iii*) The sample that gives a detector reading of 40.3 will be equivalent to a sample mass of 0.65 ng. The second sample gives a response that lies outside the linear range of detector response. If it is injected without dilution, then the results will be invalid.

Once the pre-requisites for a quantitative determination have been established, we may proceed.

****

SAQ 4.2b

The following data were obtained for a solution of benzene of unknown concentration and a series of standards. An equivalent volume of spectroscopic grade methylbenzene was added as an internal standard to the unknown and the standards. Fig. 4.2e. shows a chromatogram of one of the standards.

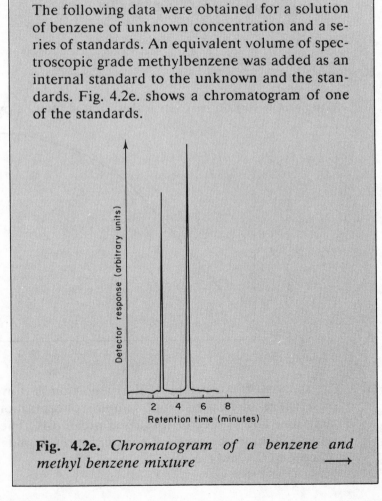

Fig. 4.2e. *Chromatogram of a benzene and methyl benzene mixture* ⟶

SAQ 4.2b
(cont.)

(*i*)   Does methylbenzene fulfil the criteria necessary to be used as an internal standard for this analysis?

(*ii*) Draw a calibration curve for the benzene standards using the internal standard.

(*iii*) Calculate the concentration of benzene in the unknown.

concentration of benzene mg dm^{-3}	peak area of benzene	peak area of methylbenzene
2.00	0.504	1.24
4.00	0.969	1.29
6.00	1.45	1.29
8.00	1.93	1.22
10.00	2.47	1.26
unknown	1.34	1.28

Response

(*i*) The internal standard does meet the following criteria:

(*a*) to be structurally similar to the compounds of interest.

(*b*) to be completely resolved from other component peaks.

(*c*) to elute close to the compound of interest (for a multicomponent mixture, it may be necessary to have a series of standards).

(*d*) to be stable and of high purity.

(*ii*) The calibration curve is shown below:

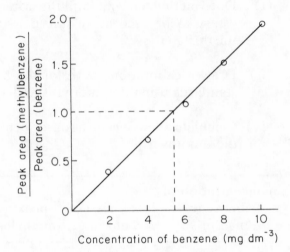

(*iii*) The concentration of benzene in the standard is 5.40 mg dm^{-3}.

****

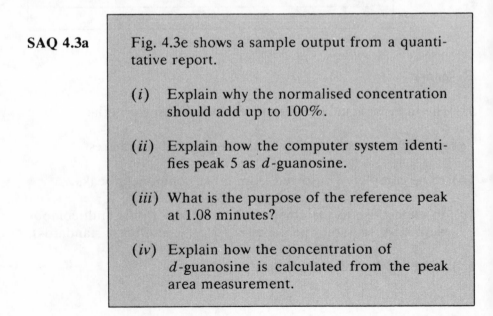

SAQ 4.3a   Fig. 4.3e shows a sample output from a quantitative report.

(*i*)   Explain why the normalised concentration should add up to 100%.

(*ii*)   Explain how the computer system identifies peak 5 as *d*-guanosine.

(*iii*) What is the purpose of the reference peak at 1.08 minutes?

(*iv*)   Explain how the concentration of *d*-guanosine is calculated from the peak area measurement.

Response

(*i*)   You may remember from Section 4.2.6 that internal normalisation is a calibration method that is based on measurement
of the total area of every peak in the chromatogram. The areas are then summed and the total normalised to 100%. Each
peak is then calculated as a percentage of the total peak area.
By definition, therefore, the total must equal 100%.

(*ii*) The computer identifies the component by the same method
that you would use; by comparing retention times of a component to that recorded by standards and held in the data bank.
In this case, retention times relative to peak 1 are recorded.

(*iii*) The reference peak is just that. It acts as a reference for recording relative retention times or as an internal standard.

(*iv*) Quantitative analysis is achieved by combining the methods of
internal and external standard. The peak area of d-guanosine
is measured and ratioed to the internal standard. This corrects
the area for variations in sample injection volume. These adjusted areas are then compared to calibration graphs obtained
previously.

****

SAQ 5.1a   Prior to performing a preparative separation using column chromatography, thin-layer chromatography is often used to check the suitability
of a particular mobile and stationary phase combination.

(*i*) Suggest a reason why this procedure is performed in preference to using column chromatography in the first instance.

(*ii*) Why do you think a less polar solvent is
often preferred to the chosen tlc solvent for
the final column separation?

Response

Separations using tlc can be achieved far more quickly than using
an equivalent column. Most tlc runs can be completed within 30
minutes, whereas the same separation by column chromatography
can take a matter of hours. A preliminary set of tlc separations
can be used to check combinations of stationary and mobile phases
quickly and effectively.

In ascending chromatography, the separation ceases as soon as the
mobile phase front arrives at the top of the plate. However, solutes
can be eluted from the end of the column; they can therefore travel
more slowly, components can be more effectively separated and still
elute from the column in a reasonable amount of time.

****

SAQ 5.1b	Before we start to discuss the types of stationary and mobile phases used in column chromatography, try to remember the four types of sorption mechanism that provide us with the basis of chromatographic separations. Describe the nature of the stationary phase and how separation of different solutes proceeds. Avoid referring to the beginning of this Unit if possible.

Response

Perhaps the first mechanism that springs to mind is *adsorption* since it is encountered in physical and other branches of chemistry. The technique is often known as *liquid-solid* chromatography. The stationary phase is a solid onto which solutes are *adsorbed*. The rate at which they will move down a column will depend on the affinity of each particular solute for the stationary phase.

A second mechanism you may have thought of uses immiscible liquids for the stationary and mobile phases and is called *partition* or *liquid-liquid* chromatography. The liquid stationary phase is coated onto the surface of a solid support. Separation depends on the differential solubility of solutes between the mobile and the stationary phases.

Ion-exchange is a technique that has certain specific applications in the separation of charged species. Separation in this case is based on the relative attraction of charged sites on the ion-exchange material for ionic species travelling in solution.

The final mechanism of the four is known under a variety of names, but *size exclusion* chromatography is the term we prefer. The column packing is made from a material that has well defined pore sizes. Smaller molecules can penetrate the pores and will thus tend to be retained on the stationary phase, while larger molecules which are too big to enter the pores move quickly down the column.

****

SAQ 5.1c	Three aromatic hydrocarbons are to be separated on a highly polar alumina column.
	Why is a non-polar solvent such as hexane a more appropriate mobile phase than propanone?

Response

The three aromatic hydrocarbons are relatively non-polar. In the presence of a more polar solvent such as propanone, there would be competition between both solute and solvent molecules for the adsorption sites on the surface of the stationary phase, which the more polar propanone would easily win. The result would be that the sample would move rapidly through the column without a separation being achieved.

It is also worthwhile considering the converse of this situation. If a non-polar solvent were used with a polar sample mixture, the mixture would tend to remain at the top of the column and again there would be no separation.

****

SAQ 5.1d	Assume that you need to separate a four component mixture and that you are provided with a column filled with an appropriate stationary phase. After elution with cyclohexane, two components of the mixture separate satisfactorily and are eluted from the column. However, the other two components stay firmly adsorbed on the top of the column. What is your course of action?

Response

You should try a slightly more polar solvent first, possibly methylbenzene which, as can be seen from Fig. 5.le, is a higher polarity
solvent and is less toxic than benzene. If the substances still failed
to elute, then progressively more polar solvents could be used until
separation was successful.

Note that we suggested that a *slightly* more polar solvent should
be used. If a highly polar solvent were to be used initially, it is
possible that it would cause the materials to co-elute, ie elute from
the column at the same time, resulting in the separation having to
be repeated.

****

SAQ 5.2a	Explain the terms 'linear capacity' and 'theoret-ical plate'.

Response

Linear capacity is defined as the amount of sample per gram of
adsorbent that just causes a significant departure of the isotherm
from linearity.

The term theoretical plate is defined as that length (height) of column within which a single equilibration is considered to occur (Sections 1.4, 3.4 and 3.5).

****

SAQ 5.2b   0.5462 g of dry ion-exchange resin in the sulphonic acid form were transferred to a 100 cm^3
beaker with 100cm^3 of deionised water. An excess, approximately 2g, of Analar sodium chloride, was added, allowed to equilibrate with the
resin and then the total volume titrated against
standard, approximately 0.1 mol dm^{-3}, sodium
hydroxide solution using screened methyl orange as indicator.

As the base was added, the indicator changed
colour at what appeared to be the end-point of
the reaction. However, after a short time, the
original colour of the indicator was regenerated.
Additional volumes of base were necessary to
ensure that the end-point reached was stable.

Total volume of base added   = 24.50 cm^3

Molarity of base   = 0.1051 mol dm^{-3}

You may use these results to answer the following:

(*i*)   Calculate the capacity of the cation-exchange
resin.

(*ii*) Suggest why several end-points were observed, before the end-point colour remained stable.

Response

(*i*) The capacity of the ion-exchange resin may be calculated as
follows:

The sodium ions replace the hydrogen ions on the resin and introduce a free acidity into the solution. The number of moles of

base that have been delivered at the stoichiometric end-point will be equivalent to the number of moles of hydrogen ion displaced.

$$\text{Number of moles of base} = \frac{24.50 \times 0.1051}{1000}$$

$$= 2.57 \times 10^{-3}$$

The dry weight capacity is defined as the amount of exchangeable ion per gram of dry ion-exchange resin.

$$\text{Thus, the dry weight capacity} = (2.57 \times 10^{-3})/0.5462$$

$$= 4.71 \times 10^{-3} \text{ mol g}^{-1}$$

(*ii*) As for the end-points, you should remember that ion-exchange is an equilibrium process, and because of the porous nature of resins a finite time will be required for exchange to reach a limiting value. This is because of the time needed for the Na^+ ions to diffuse into various pores to reach the exchange sites. Remember also that some sites may be impossible to reach.

****

SAQ 5.2c

The first three fractions, called A, B and C respectively, separated from a sample of egg white on a carboxymethyl cellulose ion-exchange column had retention volumes as follows:

A   22 cm^3
B   48 cm^3
C   76 cm^3

Given that the volume of the ion-exchanger in the column is 10 cm^3 and that the void volume is 5 cm^3, calculate the distribution coefficient of each of the components A, B and C.

Response

For sample A,

$$V_R = 22 \text{ cm}^3$$

$$V_S = 10 \text{ cm}^3$$

$$V_m = 5 \text{ cm}^3$$

By substitution in equation 5.5,

$$22 = 5 + 10\,(K)$$

hence, $$K_A = 1.7$$

Similarly $K_B = 4.3$ and $K_C = 7.1$

****

SAQ 5.2d   After fractions A, B and C, described in SAQ 5.2c were eluted, the pH of the buffer was increased from its original value of 4.8 to elute other components, at the same time increasing the ionic strength. Discuss the effect of this increasing pH and ionic strength on ions held by the stationary phase.

Response

Increasing the ionic strength will reduce the affinity of proteins for the surface due to competition between the proteins and the ions present in the buffer.

If the pH is increased, then the affinity of the proteins for the cation-exchange resin will be decreased and they will tend to move into the mobile phase.

****

SAQ 5.2e	A molecular species of relative molecular mass 120 000, which was eluted through a stationary phase of total exclusion limit 80 000, had a retention volume of 25 cm³.
	A material of low relative molecular mass had a retention volume of 225 cm³, and a sample of unknown relative molecular mass had a retention volume of 125 cm³.
	Calculate the void volume, stationary phase volume and the distribution coefficient for the sample component of unknown relative molecular mass.

Response

Void volume, $V_m = 25$ cm³, pore volume $V_s = 200$ cm³

Distribution coefficient $= 0.5$

These values may be calculated by substituting in Eq. 5.5.

The material of high relative molecular mass is not retained by the stationary phase, hence $K = 0$, and the retention volume of 25 cm³ yields directly the void volume, V_m.

The material of low relative molecular mass permeates the pores totally and will have a value of unity for the distribution coefficient,

hence, $$V_s = V_R - V_m$$

and substituting for V_M and V_R,

$$V_s = 225 - 25 = 200$$

For the sample of unknown relative molecular mass, V_R has a value of 125 cm^3.

Rearranging the equation,

$$K = (V_R - V_m)/V_s$$

$$= 0.5$$

****

SAQ 5.3a

(*i*)   Iron(III) salts are often contaminated with iron(II). Suggest a scheme for the separation of these two ions by ion-exchange.

(*ii*) The determination of fluoride is subject to a number of interferences by metal ions. Suggest a scheme to separate metal anions from fluoride, and select an appropriate resin from a catalogue of ion-exchange materials.

(*iii*) Outline a method for detecting ions separated by ion chromatography in the absence of a conductivity detector.

Response

(*i*) You should refer to Fig. 5.3b. You will find that the distribution coefficients for iron(II) and iron(III) differ markedly in 8M hydrochloric acid. An anion-exchange resin eluted with 8M hydrochloric acid will elute iron(II) very quickly. However, iron(III) will still be strongly retained. Its elution will be achieved by lowering the molarity of the acid or simply by eluting the column with water subsequent to the elution of the iron(II).

(*ii*) This is a comparatively simple problem. Passage of the solution over a cation-exchange column in the hydrogen form will result in an effluent containing HF in the absence of other metallic species. A large number of cation-exchange resins can be used for this purpose.

(*iii*) A conventional hplc system with a non-polar bonded phase column and a uv absorption detector is a very useful method for detecting ions. The technique uses a mobile phase containing an ion that absorbs strongly in the uv region; phthalate ions are typical. As ions elute, there is a reduction in absorbance thus producing a 'negative' peak as the output. Such a system allows ion chromatography to be performed with a minimal investment in new equipment.

****

SAQ 5.3b

(*i*) Compare the primary separation mechanisms of ion-exchange and size exclusion chromatography.

(*ii*) Define the terms *total exclusion limit* and *total permeation limit* for a size exclusion stationary phase.

Response

(*i*) The mechanism of size exclusion is based on steric factors, ie where the retention of a molecule is based purely on size and shape. Small molecules can completely enter the pores of the stationary phase, are held back due to diffusion and elute from the column last. Large molecules, those with a relative molecular mass above the exclusion limit, are not held up because they are too large to enter the stationary phase; they therefore elute quickly. Those of intermediate size permeate the stationary phase to varying degrees and therefore separate as they pass down the column.

The mechanism of ion-exchange is more complex in that it depends on a number of variables such as pH or ionic strength, but separation is achieved as a result of differences in the relative affinities of various ions for an ion-exchanger.

Total exclusion limit represents that relative molecular mass above which molecules will be too large to enter the pores of the stationary phase.

Total permeation limit represents that relative molecular mass below which molecules can enter the pores of the stationary phase completely.

****

Units of Measurement

For historic reasons a number of different units of measurement have evolved to express quantity of the same thing. In the 1960s, many international scientific bodies recommended the standardisation of names and symbols and the adoption universally of a coherent set of units—the SI units (Système Internationale d'Unités)—based on the definition of five basic units: metre (m); kilogram (kg); second (s); ampere (A); mole (mol); and candela (cd).

The earlier literature references and some of the older text books, naturally use the older units. Even now many practicing scientists have not adopted the SI unit as their working unit. It is therefore necessary to know of the older units and be able to interconvert with SI units.

In this series of texts SI units are used as standard practice. However in areas of activity where their use has not become general practice, eg biologically based laboratories, the earlier defined units are used. This is explained in the study guide to each unit.

Table 1 shows some symbols and abbreviations commonly used in analytical chemistry; Table 2 shows some of the alternative methods for expressing the values of physical quantities and the relationship to the value in SI units.

More details and definition of other units may be found in the *Manual of Symbols and Terminology for Physicochemical Quantities and Units*, Whiffen, 1979, Pergamon Press.

Table 1   *Symbols and Abbreviations Commonly used in Analytical Chemistry*

Å	Angstrom
$A_r(X)$	relative atomic mass of X
A	ampere
E or U	energy
G	Gibbs free energy (function)
H	enthalpy
J	joule
K	kelvin ($273.15 + t\ °C$)
K	equilibrium constant (with subscripts p, c, therm etc.)
K_a, K_b	acid and base ionisation constants
$M_r(X)$	relative molecular mass of X
N	newton (SI unit of force)
P	total pressure
s	standard deviation
T	temperature/K
V	volume
V	volt ($J\ A^{-1}\ s^{-1}$)
$a, a(A)$	activity, activity of A
c	concentration/ mol dm^{-3}
e	electron
g	gramme
i	current
s	second
t	temperature / °C
bp	boiling point
fp	freezing point
mp	melting point
$\approx$	approximately equal to
$<$	less than
$>$	greater than
e, $\exp(x)$	exponential of x
$\ln x$	natural logarithm of x; $\ln x = 2.303 \log x$
$\log x$	common logarithm of x to base 10

Table 2 *Alternative Methods of Expressing Various Physical Quantities*

1. **Mass (SI unit : kg)**

    $$g = 10^{-3} \text{ kg}$$
    $$mg = 10^{-3} \text{ g} = 10^{-6} \text{ kg}$$
    $$\mu g = 10^{-6} \text{ g} = 10^{-9} \text{ kg}$$

2. **Length (SI unit : m)**

    $$cm = 10^{-2} \text{ m}$$
    $$\text{Å} = 10^{-10} \text{ m}$$
    $$nm = 10^{-9} \text{ m} = 10\text{Å}$$
    $$pm = 10^{-12} \text{ m} = 10^{-2} \text{ Å}$$

3. **Volume (SI unit : m³)**

    $$l = dm^3 = 10^{-3} \text{ m}^3$$
    $$ml = cm^3 = 10^{-6} \text{ m}^3$$
    $$\mu l = 10^{-3} \text{ cm}^3$$

4. **Concentration (SI units : mol m^{-3})**

    $$M = \text{mol } l^{-1} = \text{mol dm}^{-3} = 10^3 \text{ mol m}^{-3}$$
    $$mg \text{ } l^{-1} = \mu g \text{ cm}^{-3} = ppm = 10^{-3} \text{ g dm}^{-3}$$
    $$\mu g \text{ } g^{-1} = ppm = 10^{-6} \text{ g g}^{-1}$$
    $$ng \text{ cm}^{-3} = 10^{-6} \text{ g dm}^{-3}$$
    $$ng \text{ dm}^{-3} = pg \text{ cm}^{-3}$$
    $$pg \text{ } g^{-1} = ppb = 10^{-12} \text{ g g}^{-1}$$
    $$mg\% = 10^{-2} \text{ g dm}^{-3}$$
    $$\mu g\% = 10^{-5} \text{ g dm}^{-3}$$

5. **Pressure (SI unit : N m^{-2} = kg m^{-1} s^{-2})**

    $$Pa = Nm^{-2}$$
    $$atmos = 101 \text{ } 325 \text{ N m}^{-2}$$
    $$bar = 10^5 \text{ N m}^{-2}$$
    $$torr = mmHg = 133.322 \text{ N m}^{-2}$$

6. **Energy (SI unit : J = kg m^2 s^{-2})**

    $$cal = 4.184 \text{ J}$$
    $$erg = 10^{-7} \text{ J}$$
    $$eV = 1.602 \times 10^{-19} \text{ J}$$

Table 3   *Prefixes for SI Units*

Fraction	Prefix	Symbol
10^{-1}	deci	d
10^{-2}	centi	c
10^{-3}	milli	m
10^{-6}	micro	μ
10^{-9}	nano	n
10^{-12}	pico	p
10^{-15}	femto	f
10^{-18}	atto	a

Multiple	Prefix	Symbol
10	deka	da
10^2	hecto	h
10^3	kilo	k
10^6	mega	M
10^9	giga	G
10^{12}	tera	T
10^{15}	peta	P
10^{18}	exa	E

Table 4 *Recommended Values of Physical Constants*

Physical constant	Symbol	Value
acceleration due to gravity	g	9.81 m s^{-2}
Avogadro constant	N_A	$6.022\ 05 \times 10^{23} \text{ mol}^{-1}$
Boltzmann constant	k	$1.380\ 66 \times 10^{-23} \text{ J K}^{-1}$
charge to mass ratio	e/m	$1.758\ 796 \times 10^{11} \text{ C kg}^{-1}$
electronic charge	e	$1.602\ 19 \times 10^{-19} \text{ C}$
Faraday constant	F	$9.648\ 46 \times 10^{4} \text{ C mol}^{-1}$
gas constant	R	$8.314 \text{ J K}^{-1} \text{ mol}^{-1}$
'ice-point' temperature	T_{ice}	$273.150 \text{ K exactly}$
molar volume of ideal gas (stp)	V_m	$2.241\ 38 \times 10^{-2} \text{ m}^3 \text{ mol}^{-1}$
permittivity of a vacuum	ϵ_0	$8.854\ 188 \times 10^{-12} \text{ kg}^{-1} \text{ m}^{-3} \text{ s}^4 \text{ A}^2 \text{ (F m}^{-1})$
Planck constant	h	$6.626\ 2 \times 10^{-34} \text{ J s}$
standard atmosphere pressure	p	$101\ 325 \text{ N m}^{-2} \text{ exactly}$
atomic mass unit	m_u	$1.660\ 566 \times 10^{-27} \text{ kg}$
speed of light in a vacuum	c	$2.997\ 925 \times 10^{8} \text{ m s}^{-1}$